BASIC FLUID SYSTEM ANALYSIS

by

G. A. Patterson

Basic Science Press
1608 Via Lazo
Palos Verdes Estates
California, 90274

Library of Congress Number — 76-21585
ISBN — O-917410-00-9

PREFACE

The aim of art is to please the senses; the aim of science is to understand the relations between natural laws; and the aim of engineering is to transport and transform materials and energy. Engineering technology in the fields of transportation, power, agriculture and chemistry is based on the usage of liquids and gases which comprise the alternate states of fluids.

Fluids are stored in tanks and reservoirs, transported through pipes, caused to move by pumps and controlled by valves and regulators. The design of fluid systems is a specialized art which requires individualized methods and data for each application. There is a need for different considerations in the liquid hydrogen system, the irrigation system, and the steam system. There are however, similar elements to all of these applications. The present work provides simplified procedures for flow analysis which are applicable to all systems.

CONTENTS

LIST OF TABLES

LIST OF FIGURES

INTRODUCTION

Fluid system analysis deals primarily with determining flow rates and pressures in liquid and gaseous systems which store and transport fluids. The present work provides simplified mathematical and graphical methods for making these calculations.

The computations set forth in this book can be performed step-by-step. In addition, the same computations are arranged in Appendix I as programs for a programmable pocket calculator.

The programmable pocket calculator is a quantum jump above the electronic slide rule for the reasons that (1) all steps of the program can be entered simultaneously and a sample validity check performed, and (2) the program can be automatically repeated with incremental changes approaching natural growth and decay processes. In addition, calculator programs free the mind from repetitive elemental reasoning allowing the solution to become more complex and representative.

The analyst has the additional responsibility to make provisions for system safety and satisfactory operation over the required range of temperatures and other variables. This can be done by use of valves, regulators, or other devices, or by use of proper procedures. Basic considerations are discussed with particulars to be established by the specific application.

2.0 FLUID SYSTEM DESIGN ELEMENTS

2.1 Fluid Motion

2.1.1 Forms of Pressure

Bernoulli laid the foundation for fluid analysis in 1740 by relating total pressure, static pressure, and velocity pressure. It is a simple relationship. Total pressure equals the static pressure plus the velocity pressure. Static pressure equals the total pressure minus the velocity pressure.

$$P_t = P_s + P_v$$

where P_t = Total Pressure, lb/ft^2
P_s = Static Pressure, lb/ft^2
P_v = Velocity Pressure, lb/ft^2

Velocity pressure was quantified by Henri Pitot at the same time as Bernoulli's studies. Pitot dipped a bent glass tube into the River Seine and found the water rose in the tube to a height (the "head") proportional to the square of the velocity of the river. This observation is reflected by the formula:

$$V = (2gH)^{1/2}$$

or $$H = V^2/(2g)$$

where V = Velocity, ft/sec
H = Head of fluid, feet
g = Gravity force, 32.2 ft/sec^2

Note that the above equation for H can be rewritten to obtain the velocity pressure, Pv or "q".

$$H\,(g\rho) = (1/2)\,((g\rho/g)\,V^2$$

or $$P_v = (1/2)\,\rho\,V^2$$

where $g\rho$ = Specific weight of fluid, lb/ft^3
ρ = Density of fluid, $(lb/ft^3)/(32.2\ ft/sec^2)$

2.1.2 Flow Measurement

Torricelli in 1640 measured the flowrate from a hole in a cask and found the flow to vary with the square root of the liquid head. This observation has a close similarity to the later work of Pitot who found the velocity of flow to be proportional to the square root of the head. This follows since for a constant flow area the flow rate is proportional to the velocity of the fluid.

The volume flow can be found by knowing the flow area and the velocity.

$$dQ = A\,V$$

where dQ = Volumetric flow rate, ft^3/sec

A = Area, ft^2

V = Velocity, ft/sec

The weight flow rate is found by multiplying the volume flow rate, dQ, by the specific weight

$$dW = dQ\,g\rho$$

where dW = Weight flow rate, lb/sec

$g\rho$ = Specific weight, lb/ft^3

The geometric area may not be fully effective as a flow area. The formation of vortices prohibits the flow path from following excessively abrupt geometric changes so that a "discharge coefficient" is required. This coefficient can range from 0.6 for sharp edged holes to nearly 1.0 for well rounded inlets.

Orifice plates are the simplest means of measuring flow. Static pressures are measured upstream and downstream near the plate to determine velocity. Pressure drop is high but the plates are usually simple and cheap to install.

Weirs are partly submerged orifices used in irrigation ditches and open water supply systems. Weirs may have a decreasing area at the bottom to obtain a higher head at low flow rates.

Venturiis are named for their investigator, circa 1791, and allow flow measurement with low pressure drop. There is a gradual area reduction at the inlet to the "throat" at the minimum area followed by a gradual expansion in the diffuser to the original diameter. Venturi flow coefficients approach 1.0.

Standards for the construction and usage of flow instruments have been established by various societies, see Reference 2.

Figure 1 . Flow Measurement

1.1 Pitot Tube

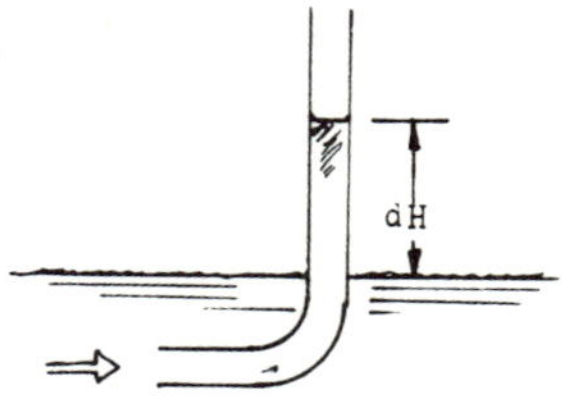

1.2 Orifice Meter

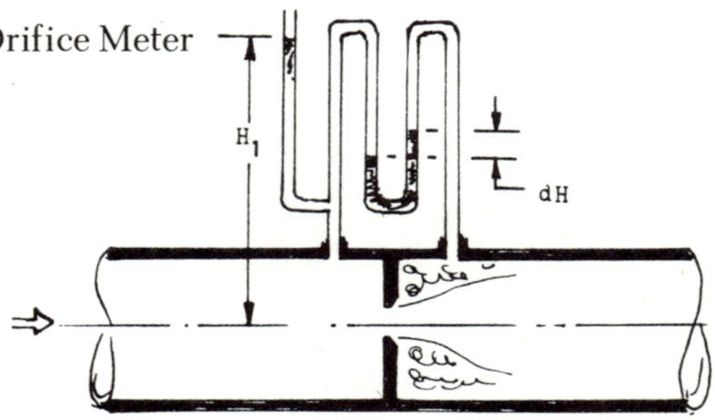

1.3 Venturi

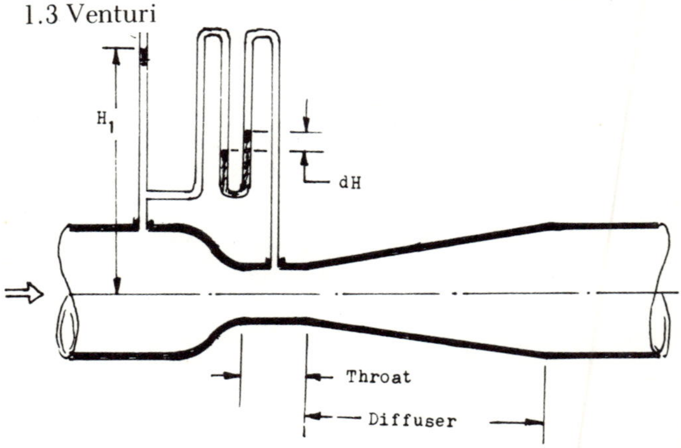

2.1.3 Limits of Velocity

There are limits to fluid velocity set by the initial conditions of pressure and temperature and according to whether the fluid is a liquid or a gas.

Liquid velocity is limited by the formation of vapor. The limiting velocity pressure "q" is nominally equal to the difference between the total pressure and the vapor pressure. The presence of dissolved gas reduces the velocity that would otherwise be obtained.

The limiting value of q for a liquid system is therefore:

$$q = P_t - P_{vapor}$$

Gas velocity is limited by the velocity of sound in the gas. At the minimum cross-sectional flow area in the system the flow rate of the gas will adjust itself to not exceed the sound velocity. This is discussed in a later section.

2.1.4 Vortex Motion

Vortex motion is circular motion. A spectrum of different types of vorticies can exist and have been classified by Stepanoff, ref.1, according to the rate at which velocity changes with the change of radius.

The liquid whirlpool or the air tornado is an example of the "free" vortex. The velocity increases toward the center until a limiting velocity is reached as determined by viscosity and friction. The velocity energy is subtracted from the total energy of the fluid at rest. This results in a lowering of the liquid level toward the center of the whirlpool and a lowering of the air pressure toward the center of the tornado.

$$V = C_3/R$$

$$H = -C_3/(2gR)$$

where
- V = Tangential Velocity, ft/sec
- R = Radius, ft
- C_3 = Constant found by measurement of V and R at one location
- H = Velocity Head, ft
- g = Gravity constant, 32.2 ft/sec^2

Another case occurs in a centrifugal pump as a "forced" vortex. Energy is added by the pump so that as the fluid is decelerated at the periphery the static pressure rises.

$$V = C_7 R$$

$$H = C_7 \; R^2/(2g)$$

A vortex can be formed where there is an abrupt change in the pressure in the fluid as in a pump inlet, at an airplane wing tip, or at the trailing end of poorly streamlined bodies. The vortex sometimes moves from one point of attachment to another giving rise to pressure oscillations.

2.1.5 Pump Performance

2.1.5.1 Centrifugal Pumps

Centrifugal pumps are considered to include geometric shapes which range from radial flow to axial flow. A parameter known as "specific speed" was devised by Camerer, Berlin, 1915, to categorize pump performance and to select pump impeller types as well as to anticipate peak efficiencies:

$$N_S = RPM\ Q^{1/2}\ H^{3/4}$$

where

N_S = Specific Speed
RPM = Rotational Speed, Revolutions/Minute
Q = Flow Rate, Gallons/Minute
H = Head Rise, Feet

Pumps having a low specific speed, in the range of 500, are most suitably radial flow and have an efficiency on the order of 50 percent, see Reference 3.

Pumps having a high specific speed, in the range of 10,000, are most suitably axial and are expected to have an efficiency of 80 to 90 percent. For intermediate specific speeds the optimum shape is a blend between the radial and axial.

Pump "affinity laws" can be used to predict performance of a pump of different size than a calibrated model or for a different speed:

$$GPM_2 = GPM_1\ (RPM_2/RPM_1)$$
$$HEAD_2 = HEAD_1\ (RPM_2/RPM_1)^2$$
$$POWER_2 = POWER_1 (RPM_2/RPM_1)^3$$

The major factor in selecting pump speed is design compatability with the driving mechanism, that is, use of minimum gearing. Synchronous alternating current motors have a speed controlled by the number of poles, usually 2 to 12. The pump speed is less than the synchronous speed by an amount established by the power extracted by the shaft.

$$\text{Synchronous RPM} = (\text{Cycles/Minute})/(\text{Number of Poles})$$

Figure 2. Centrifugal Pump Performance

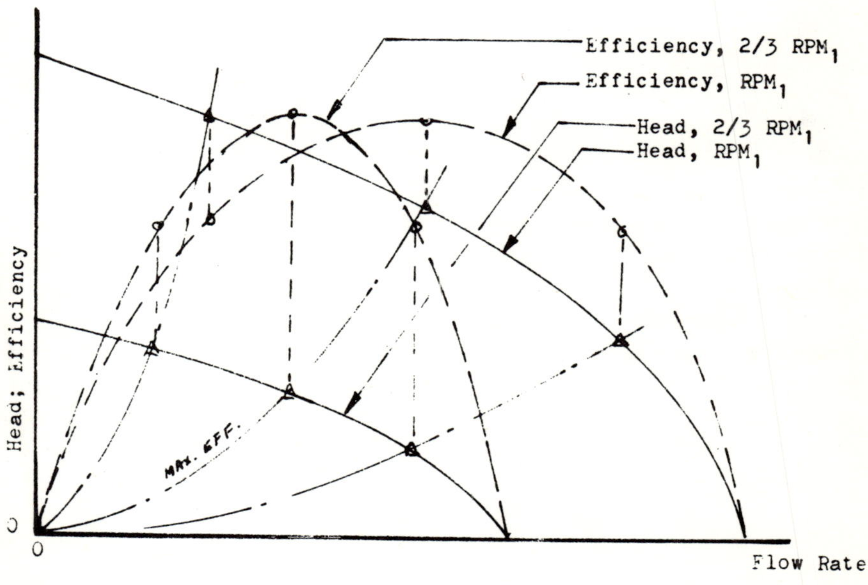

A second factor in selecting pump speed is the avoidance of vapor formation or "cavitation". Vapor can form at the pump inlet if the velocity is excessive for the fluid under the operating conditions. Vapor can also form at the low pressure areas of the impeller (the upper aft surfaces of the blades) because of improper contours, excessive flow directional changes, or mismatched blade pitch angles at the pump RPM. Conditions which affect cavitation are liquid volatility, temperature, tank pressure, inlet liquid head, pump surface finish, and inlet geometry. In aircraft tank boost pumps the aircraft climb rate is very important as the critical condition occurs at the onset of fuel boiling and climb rate affects the rate of release of dissolved air. Experimental data are required to establish the pump performance under cavitating conditions.

2.1.5.2 Positive Displacement Pumps

Positive displacement pumps include gear pumps, piston pumps, screw pumps and vane pumps. Displacement is proportional to RPM. Pressure rise is determined by the pressure drop of the system. Bypass valves are required to allow changes to the system flow rate. Systems having bypass valves incur temperature buildup since energy is dissipated. A bypass system is susceptible to cavitation at low flow rates because of this temperature increase. Positive displacement pumps are usually certified to operate at a fixed "pressure above vapor pressure" for normal operation to avoid cavitation.

2.1.5.3 Jet Pumps

Jet pumps depend upon bleed motive flow from another source. They are simple to construct and operate but require analysis to define performance for a specific application. Performance can be estimated by the methods given later according to data and procedures of Cunningham, Reference 3.

2.2 OPERATIONAL PRECAUTIONS

2.2.1 Contaminates

A fluid system must be designed to operate with contaminates of a specified type and quantity for a specified time before maintenance. Contaminates can include other liquids and gases including water and air and acids as well as microbes, algae, and solid particles of all sizes and materials. Fibers can also be present.

Microbes can produce acids which react with the structure of the tank or lines. The presence of water encourages the presence of microbes.

Liquids normally contain water condensed from the atmosphere. Most liquids dissolve water as a function of temperature; lower temperatures reduce the solubility causing water to be released from solution. At slow cooling rates the water is released as small droplets which below 32 degrees form snow crystals; at fast cooling rates the water is released at a delayed rate, may not be released until cooled below 32 degrees and agitated, and may form glaze ice crystals which can quickly plug filters.

2.2.2 Thermal Expansion

Liquid systems must be designed to accommodate an increase in volume according to the design operating temperature range and the expansion characteristics of the liquid. Expansion must be considered both in the supply tank and in lines which contain liquid locked between closed valves.

Design accommodation for the tank usually consists of having a fill level sufficiently below the overflow level. For the lines it consists of having relief valves which can prevent excessive pressure that would cause structural failure.

The line itself or its supports may change dimensions as a result of changes in temperature or pressure. This can require expansion joints as well as line constraints.

Thermal expansion can cause siphoning of fuel from a tank where the vent line outlet is below the liquid level in the tank. This is particularly a problem in aircraft tanks and air inlets at the high point of the line, "siphon breakers," are added to avoid excessive negative tank pressure.

2.2.3 Hazards

The handling of fluids can be hazardous. A design safety review should be made for each new fluid system. Applicable design codes and regulations must be followed.

Precautions should be taken to prevent and control damage to equipment and personnel by reason of (1) exposure to toxic materials, (2) exposure to high and low temperature liquids and gases, (3) contact with high velocity liquids and gases, (4) excessive noise and vibration, (5) debris of explosions caused by structural failure, and (6) the consequences of combustion and detonation.

High pressure gases possess an elastic quality which tends to produce a damage range of greater radius than a similar liquid system failure. For this reason, as well as the greater ease of developing pressure, liquids are favored for testing high pressure vessels.

Combustion precautions should be taken against ignition by (1) lightning, (2) exposed sparks of electrical equipment, (3) excessive surface temperatures, and (4) frictional electrification of liquids and gases. In addition, some fluids are capable of being detonated on impact.

Frictional electrification of gases is an insidious danger. There are cases of explosions caused by addition of steam for cleaning and addition of carbon dioxide for inerting. Sparks were released between droplets which had condensed and become electrified to ignite the combustible mixture present in the tank. Filling a fuel tank from above at high velocity can produce similar sparks and cause explosions.

Frictional electrification of particles in liquids can also occur. This is ordinarily harmless, however, explosions can result by electrodes present in the vent area which connect tank and liquid charges.

2.3 SYSTEM DESIGN APPROACH

2.3.1 General Analysis Approach

Before proceeding with the analysis it is desirable to determine which phenomena are significant. For example, if a tank is to be emptied or filled is the internal flow restriction of importance? If the tank contains baffles of significant head loss at the required external flow it will be necessary to consider these internal flow restrictions.

A sketch should be made to illustrate the anticipated findings in terms of "available flow" versus "required flow" or in similar applicable terms. Graphs should have zero-zero ordinates to facilitate interpolations and extrapolations and to keep the values in proper perspective. The pressures and flow limits should be indicated as well as the expected variation in values according to geometric concepts. For example, line pressure drop is increased by a factor of four as the flow is doubled. This is the form of a parabola and can be readily sketched on a study graph. These drawings can greatly simplify the problem, promote confidence, and indicate errors of computation or of concept if unexpected answers are obtained. It is usually best not to proceed until the answers have been anticipated in their general nature.

2.3.2 Dimensional Analysis

Engineering calculations require assembling various equations and data having diverse origins to form a coherent unity useful to the application. Rules for this process were formulated by Buckingham and by Lord Rayleigh in 1915. The process is known as dimensional analysis, see Reference 4.

Dimensional analysis tells us to equate apples to apples. More than that, we must equate complex units, as gallons per minute, to each other. When we combine different numbers and different units we get a product of both, as 24 inches times (1 foot/12 inches) equals 2 feet.

Note in the above example that the dimensions and the values above the line cancel those below the line. Similarly, dimensions on opposite sides of the equal sign cancel each other if of the same exponent. Exponents above the line cancel those below the line for the same type of dimensions.

Examples: $\text{feet}^2 \times \text{foot} = \text{feet}^3$

$\text{feet}^4 \times \text{foot} = \text{feet}^5$

$\text{feet}^6 \times \text{feet}^2 = \text{feet}^8$

$\text{feet}^6/\text{feet}^2 = \text{feet}^4$

Exponents on the same side of the line add; those on opposite sides subtract. Exponents are frequently written as negative values for convenience:

$\text{feet}^{-3} = {}^1/\text{feet}^3$

Exponents which are to be cubed or squared are actually multiplied. They are divided to find the square or cube root:

$$(\text{feet}^2)^2 = \text{feet}^4$$
$$(\text{feet}^7)^2 = \text{feet}^{14}$$
$$(\text{feet}^3)^{1/2} = \text{feet}^{3/2} = \text{feet}^{1.5}$$
$$(\text{feet}^9)^{1/3} = \text{feet}^3$$

Groups of dimensions peculiar to a physical phenomenum and which form a dimensionless number have great significance in science and engineering. Particularly useful to fluid studies are the Reynolds Number and the Mach Number, after Ernst Mach, circa 1900:

$$RN = D\,V\,\rho/\mu$$

where
- D = Diameter, feet
- V = Velocity, feet/sec
- ρ = Mass Density, $(\text{lb/ft}^3)/32.2\ \text{ft/sec}^2$
- μ = Viscosity, lb-sec/ft^2

$$MN = V/Vc$$

where Vc = Sound Velocity, ft/sec

2.3.3 The Use of Schematics

A schematic or pictorial is essential for the design and analysis of a fluid system. The schematic should include components, flow paths, sizes, elevations, pressures and other data as necessary to describe the system. Ordinarily, only a portion of the total installation is being studied and it is useful to extract a partial schematic.

The flow paths of a proposed system should be critically examined to determine if the functions of the system (filling, emptying, transfer, and others) can be accomplished. Each function should be studied in its entirety to identify the lines used, valve positions, flow paths, and compatibility with other system functions. Component failure should be considered (open, closed, and structural) for effect on safety and reliability. The failure study may indicate a need for system arrangement changes or for redundant components or for check valves to prevent reverse flow.

2.3.4 Flow System Arrangements

A flow system can include parallel lines (one alongside the other) as well as series lines (one following the other). In parallel lines which are connected at both ends the pressure drop is the same in both lines and the flow is distributed by the ratio of the pressure drop as necessary to achieve this pressure equality. In parallel lines which are connected at only one end the pressure is the same at the junction and the flow is found by knowing the pressure at the other ends. In series flow the pressure drop is the total for all lines and the flow is the same in all lines. Frequently, both flow and pressure drop are unknown and it is necessary to select trial values and to match potential flow to potential losses. Graphical presentations are highly desirable to reduce the amount of computation and the complexity of mathematical analysis.

2.3.5 Performance Predictions

The accuracy with which the system performance can be predicted depends not only on the calculation methods employed but by the precision with which the system can be defined. During preliminary design there is uncertainty in final physical arrangement, in sizes, in elevations, and in the performance of components. There must be allowed a tolerance for manufacturing dimensions which affect pressure rise and flow rates. During operational life there may be system deterioration, contaminant buildup, and operation beyond the design values. In addition, there may be uncertainty in flow coefficients and in fluid properties. The analyst must make reasonable selections of methods and data to suit his knowledge of the application.

The successful analyst must not only develop proficiency in the elements of flow analysis but must constantly review the scope of the analysis to determine if it should be enlarged to include other phenomena. For example, is surge a consideration?, is thermal expansion?, is vaporization a possible problem?

Engineering makes use of many approximations, both in terms of data and in terms of mathematical relations. To be able to identify the key parameters, the critical time of an event, and the magnitude of a correct answer is the very essence of successful engineering. But the problem changes; the accuracy demanded of one case may be grossly excessive or inadequate for another. Methods and data should be critically examined before being applied to a specific case.

3.0 THE ELEMENTS OF ANALYSIS

3.1 LIQUID FLOW

Flow through a conduit occurs at a rate at which the power input equals the system losses or the pressure drop. If the flow is caused only by an elevation difference or "head" the power input and the pressure drop is equal to the difference in elevation measured at the initial and final liquid level times the specific weight of the liquid. If the outlet is not submerged the elevation is measured from the outlet. If a tank is being emptied or filled the head difference will change with the level, and successive calculations or an integral solution are necessary to describe the process. If there is a pump or tank pressurization these values must be added to the elevation difference to obtain the total pressure causing flow. In any case, the horsepower expended is as follows:

$$HP = \frac{(lb/ft^2) \times (lb/sec)}{(lb/ft^3) \times 550\ (ft\text{-}lb/sec)/HP}$$

Pressure drop is the summation of energy losses caused by friction along the boundary surface plus acceleration and deceleration losses caused by directional changes of the flow path which can produce energy consuming vorticies. These are computed separately as "line losses" and as "component losses".

Component losses under turbulent conditions are substantially independent of Reynolds Number and can be correlated according to the number of velocity heads lost, being known as the "k" value. Typical k values are given in Tables I through V for fittings, bends, inlets, miter joints, and expansions.

Line friction loss is computed according to the friction factor, the length/diameter ratio and the velocity pressure, as

$$dP = f\,(L/D)\,q$$

The friction factor "f" is computed by several equations given below according to whether the flow is laminar or turbulent and if turbulent whether the pipe is smooth or of varying degrees of roughness.

Under laminar flow the liquid particles move in streamlines so that an injection of a dye upstream will form a continuous unmixed trace at downstream locations. This occurs only at low velocities as determined by Osborne Reynolds in 1883. The laminar friction factor is:

$$f = 64/RN$$

Above a Reynolds Number of 1500 to 3000 the friction factor can abruptly increase to a much higher value as the flow becomes turbulent. Ink injected into turbulent flow results in a chaotic mixture pattern. For smooth tubes the turbulent friction factor is approximately:

$$f = 0.184(RN)^{-0.20}$$

The rough tube friction factor was correlated by Nikuradse in 1932 by glueing sand grains of various sizes to the inside of pipes, Ref. 5. The rough tube fully developed (past the laminar transition) turbulent friction factor is approximately:

$$f = (1/(2 \log10(r/k) + 1.74))^2$$

where
r = radius of pipe, feet
K = diameter of surface roughness particle, feet

Figure 3. Tube Friction Factors

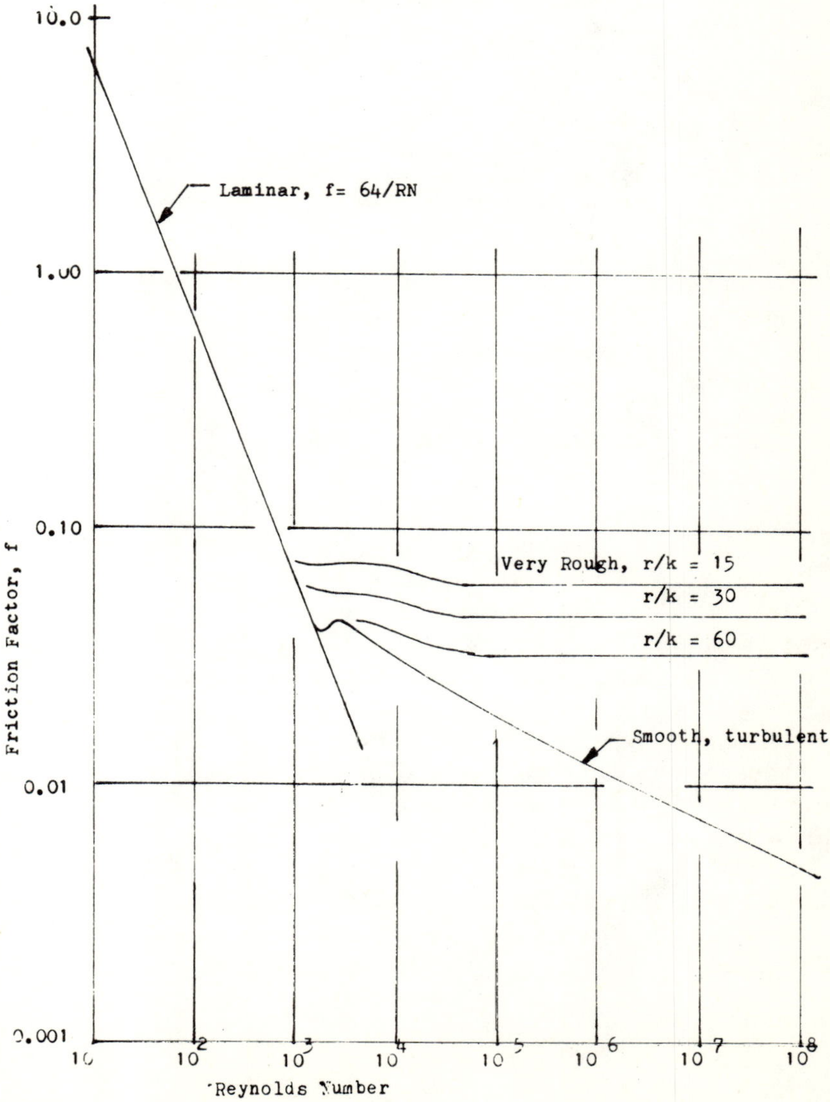

TABLE I FITTING LOSSES
(Ref. 6)

FITTING TYPE	TYPICAL K
TEE, STRAIGHT FLOW	0.3 to 0.5
TEE, ANGLE FLOW	1.0 to 1.7
GATE VALVE	0.1 to 0.3
SWING CHECK	1 to 4
ANGLE VALVE	2 to 4
GLOBE VALVE	5 to 10

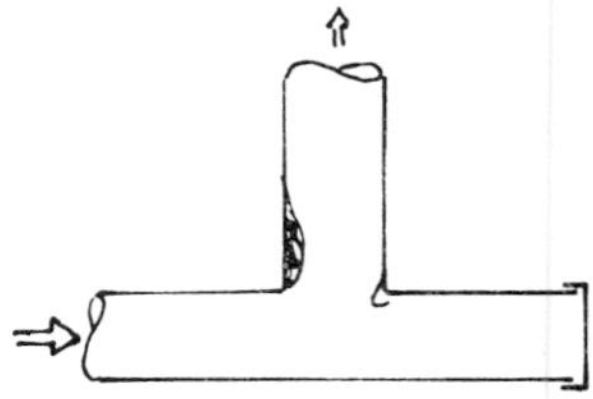

TABLE II SMOOTH BEND LOSSES
(Ref. 7)

BEND RADIUS / PIPE RADIUS	TYPICAL K 45 deg.	TYPICAL K 90 deg.	TYPICAL K 180 deg.
2	0.15	0.26	—
4	0.12	0.18	0.22
12	0.15	0.23	0.43
20	0.20	0.34	0.68
30	0.26	0.47	0.94
58	0.42	0.88	—

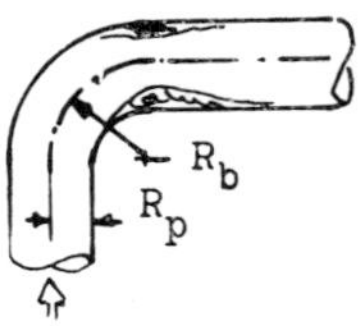

TABLE III INLET LOSSES

INLET RADIUS / INLET DIAMETER	(Ref. 6) K	(Ref. 8) K
0.00	1.00	0.46
0.02	0.62	0.30
0.04	0.40	0.17
0.06	0.25	0.10
0.08	0.14	0.04
0.16	0.00	0.00

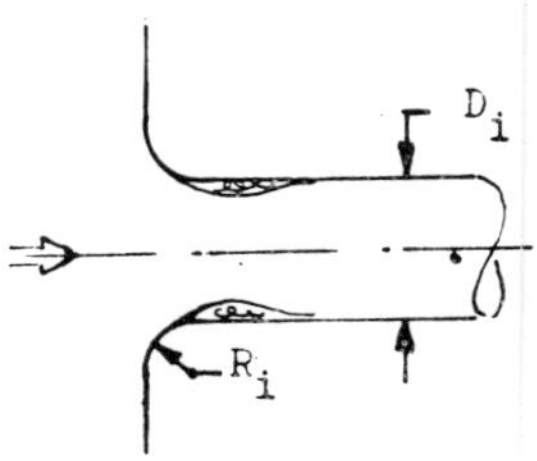

TABLE IV MITER JOINT LOSSES
(Ref. 6)

ANGLE, DEGREES	K
30	0.20
45	0.38
60	0.56
90	1.20
110	1.80

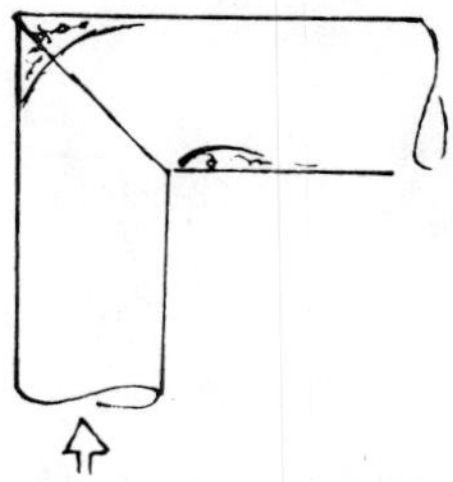

TABLE V EXPANSION LOSSES, TURBULENT FLOW
(Ref. 6)

INLET DIAMETER / EXIT DIAMETER	K
0.00	1.00
0.2	0.90
0.4	0.70
0.6	0.40
0.8	0.13
1.0	0.0

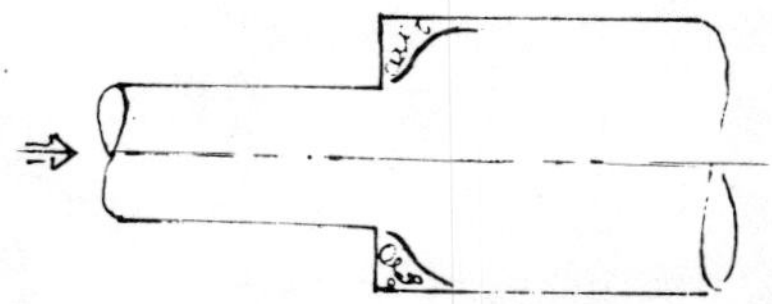

TABLE VI. LIQUID PROPERTIES
(Ref. 9)

Material	Temperature	Specific Weight	Viscosity
	Degrees F.	Lb/Ft^3	$Lb\text{-}Sec/FT^2$
Water	60	62.4	2.30×10^{-5}
	100	61.9	1.44×10^{-5}
	150	61.1	8.99×10^{-6}
	200	60.2	6.27×10^{-6}
Octane (C_8H_{10})	0	44.0	1.88×10^{-5}
	100		9.61×10^{-6}
	200		5.64×10^{-6}
	300		3.65×10^{-6}
Propane (C_3H_8)	0	31.6	3.34×10^{-6}
	100		2.09×10^{-6}
Hexane (C_6H_{14})	0	41.3	9.82×10^{-6}
	100		5.85×10^{-6}
	200		3.76×10^{-6}
	300		2.61×10^{-6}
Ethylene Glycol ($C_2H_6O_2$)	112	68.3	2.09×10^{-4}
	200		4.81×10^{-5}
	300		1.15×10^{-5}
Kerosene	-50		1.25×10^{-4}
	60	48.0	4.39×10^{-5}
	200		1.44×10^{-5}

PROCEDURE FOR LIQUID PRESSURE DROP

TURBULENT FLOW

(See Appendix I, Program Numbers 1 and 3)

1. Find Velocity

$$V = dW/(g\rho(\pi/4)D^2)$$
$$ft/sec = (lb/sec)/((lb/ft^3)(ft^2))$$

2. Find Reynolds Number

$$RN = DV\rho/\mu$$
$$= (ft)(ft/sec)(lb/ft^3/32.2\ ft/sec^2)/(lb\text{-}sec/ft^2)$$

3. Test for Turbulent Flow

Is RN more than 3000?

If yes, is turbulent, continue

Is pipe smooth?

If yes, $f = 0.184(RN)^{-0.20}$

If no, $f = (1/(2\ \log10(r/k) + 1.74))^2$

(See Figure 3)

4. Find Sum of Friction and Fitting Loss Factors

$$= f(L/D) + k$$

5. Find Dynamic Pressure

$$q = 1/2\rho V^2$$
$$lb/ft^2 = ((lb/ft^3)/32.2\ ft/sec^2)(ft/sec)^2$$

6. Find Pressure Drop

$$dp = (f(L/D) + k)\,q$$

where dp = Pressure Drop, lb/ft^2

LAMINAR FLOW

See Appendix I, Program Number 2

As above, but:

3. For RN less than 3000

$$f = 64/RN$$

Figure 4. System Diagram

Location	Length feet	Dia feet	k
A-B	100.	0.25	3.0
B-C	30.	0.15	2.5
B-D	40.	0.10	3.0

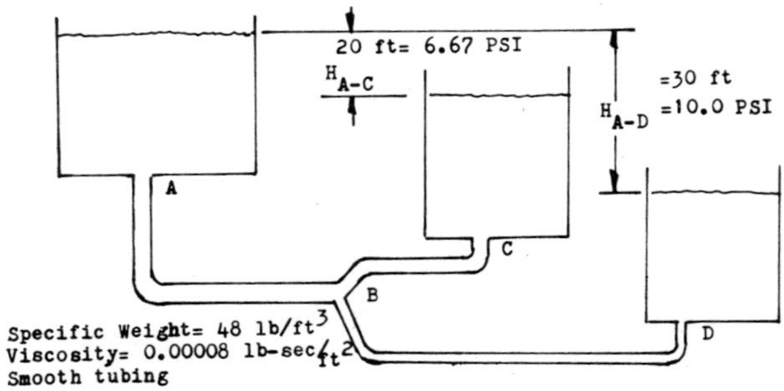

Figure 5. Calculation Results

Location	Trial Flow lb/sec	Trial dP psi
A-B	13.0	1.99
B-C	7.0	2.62
B-D	3.0	4.52

EXAMPLE OF LIQUID FLOW (ELEVATION HEAD)

Figure 6. Analysis Plot

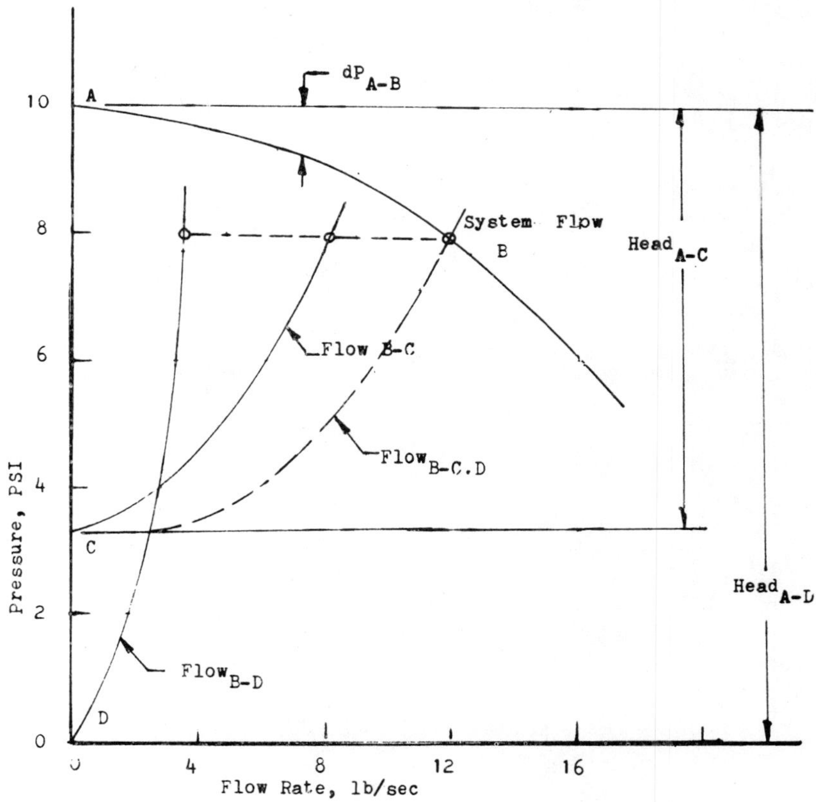

Figure 7. System Diagram

Location	Length feet	Dia feet	k
A-B	100.0	0.25	3.0
B-C	30.0	0.15	2.5
B-D	40.0	0.10	3.0

Pump Pressure
No Flow = 50.0 PSI
10.0 lb/sec = 40.PSI
16 lb/sec = 20. PSI

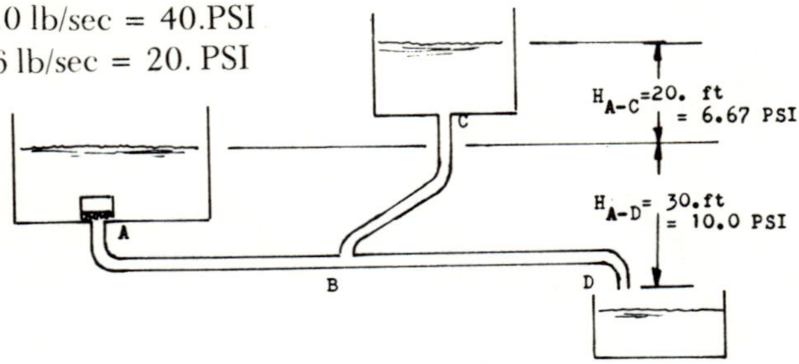

Specific Weight = 48.0 lb/ft³

Viscosity = 0.00008 lb-sec/ft²

Figure 8. Calculation Results

Location	Trial Flow lb/sec	Trial dP PSI
A-B	13.0	1.99
B-C	7.0	2.62
B-D	3.0	4.52

EXAMPLE OF LIQUID FLOW (CENTRIFUGAL PUMP)

Figure 9. Analysis Plot

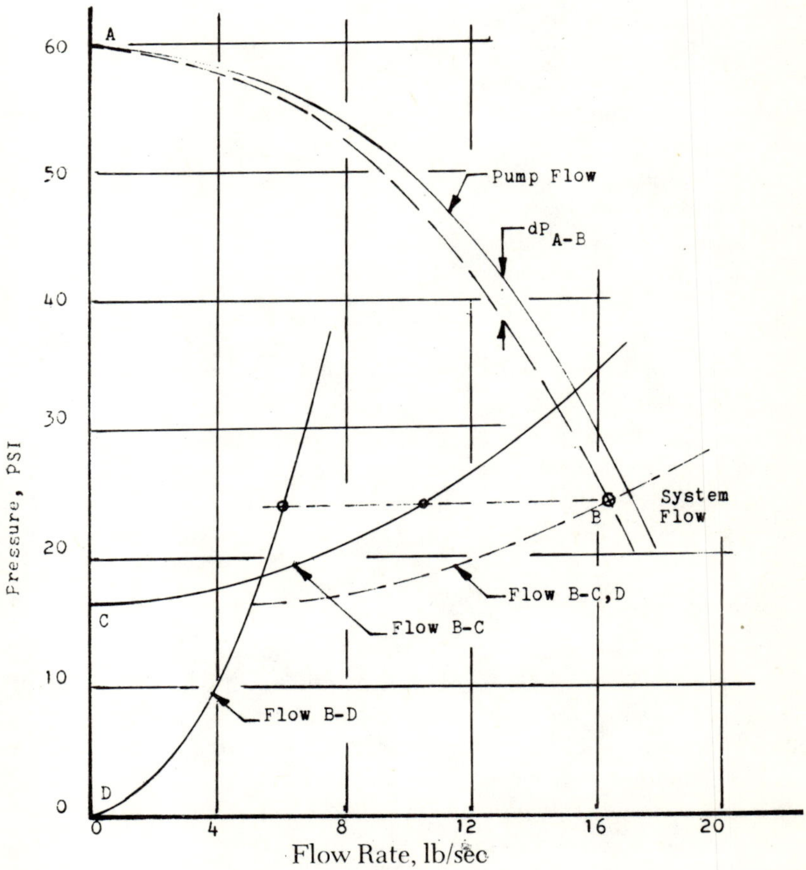

3.2 GAS FLOW

The velocity of a gas ducted through a tube or orifice cannot exceed the sonic velocity (MN = 1) at the smallest cross section. For air and most other gases this sonic velocity is reached at a pressure ratio of approximately 0.53. At a lower pressure ratio the velocity remains constant with increasing upstream pressure but the weight flow increases proportionately. At constant upstream pressure and decreasing downstream pressure the flow remains constant once the critical pressure ratio is reached. The relation between pressure ratio and the flow coefficient can be found by the following equation, see Eshbach. Reference 5.

$$\emptyset = (8.02\ R^{1/2})(k/(k-1)((P_2/P_1)^{2/k}-(P_2/P_1)^{(k+1)/k}))^{1/2}$$

where $\emptyset$ = Flow Coefficient

R = Gas Constant = 1545/Molecular Weight

k = Ratio of Specific Heats, usually = 1.4

By coincidence, the value of the flow coefficient at the sonic condition is also approximately 0.53 for air.

The orifice flow equation is simply:

$$dW = \emptyset\, C_d P_1 A / T_1^{1/2}$$

where dW = Weight flow rate, lb/sec

C_d = Shape Discharge Coefficient, 0.6 to 1.0

P_1 = Inlet total Pressure, lb/ft^2

A = Flow Area, ft^2

T_1 = Inlet total Temperature, degrees Rankine

(Note that P_1 and A can be in any consistent units)

The gas orifice formula can be used for gas line pressure drop calculations by replacing the orifice discharge coefficient with a value calculated to account for the line pressure losses. The method is given later in this section.

A simplification of the flow coefficient can be made for most computations:

$$\emptyset = 0.53\,((1-(P_2/P_1))/0.53)^{0.25}$$

Gas flow is suitably plotted on a graph of weight flow rate versus pressure with zero-zero ordinates. A straight line extending upward from the origins can be drawn to show the limit of flow at any pressure at the specified temperature and represents the sonic condition. Similarly, other lines can be drawn from the origin at lesser flow rates to represent particular fractional sonic flow conditions. At a given downstream pressure a curve can be drawn from the vertical origin at this pressure to become tangent with the sonic flow line at the upstream pressure for which P_2/P_1 equals 0.53.

Likewise, at a given upstream pressure a curve can be drawn from the vertical origin at this pressure to become tangent to a vertical line at the downstream pressure for which the pressure ratio is 0.53 and at the flow rate allowed by sonic flow at the upstream pressure.

Relief valve flow rates can be superimposed on the orifice flow curves to show composite performance. The initial flow rate is set by the opening pressure of the relief valve spring (or by area ratios for regulators of this type).

An expression for an equivalent discharge coefficient for a line can be derived as follows:

$$dP = (f(L/D) + k))\,(1/2)\,\rho V^2$$
$$V = dW/g\rho A$$
$$\rho = g\rho/g$$
$$dP = (f(L/D) + k))\,(1/2)\,(g\rho/g)\,(dW/(g\rho A))^2$$

by transposition

$$dW = ((2g\,dP)/(f(L/D) + k)g\rho)^{1/2} A\, g\rho$$
$$dW = (D/(fL + kD)^{1/2}(2g\,dP\,g\rho)^{1/2} A$$
$$dP = H\,g\rho$$
$$dW = (D/(fL + KD)^{1/2}(2gHg\rho g\rho)^{1/2} A$$
$$dW = (D/(fL + KD)^{1/2}(2gH)^{1/2} g\rho A$$

The resulting equation is seen to be the form of the simple orifice equation so that:

$$C_d = (D/(fL + kD))^{1/2}$$

TABLE VII. GAS PROPERTIES
(Ref. 5)

Gas	Ratio of Specific Heats	Gas Constant R ft/°F	Viscosity at 60°F lb-sec/ft²
Helium	1.66	386.5	4.1×10^{-7}
Air	1.40	53.35	3.8×10^{-7}
Oxygen	1.40	48.31	4.2×10^{-7}
Nitrogen	1.40	55.20	4.1×10^{-7}
Hydrogen	1.41	766.5	1.9×10^{-7}

PROCEDURE FOR GAS ORIFICE FLOW

(See Appendix I, Program Number 4)

1. Test for Sonic Flow

   If P_2/P_1 less than 0.53 flow is sonic

   If P_2/P_1 more than 0.53 flow is subsonic

2. Find Flow Factor

   Sonic: $\emptyset = 0.53$

   Subsonic: $\emptyset = 0.53\,((1-(P_2/P_1))/0.53)^{0.25}$

3. Find Flow Rate

   $dW = \emptyset\, C_d P_1 A / T_1^{1/2}$

   where dW = Flow Rate, lb/sec

   P_1 = Inlet pressure, lb/inch²

   A = Area, inch²

   C_d = Shape Coefficient, 0.6 to 1.0

   T_1 = Inlet temperature, degrees Rankine (460 + deg F)

PROCEDURE FOR EQUIVALENT GAS LINE DISCHARGE COEFFICIENT

(See Appendix I, Program Number 5)

The flow rate must be known or assumed.

1. Find Reynolds Number

   $RN = dW/(25.25D\mu)$

   dW = Flow Rate, lb/sec

   D = Diameter, feet

   μ = Viscosity, lb-sec/ft²

2. Find friction factor for turbulent flow

   $f = 0.184(RN)^{-0.20}$

3. Find discharge coefficient

   $Cd = (D/(fL + kD))^{1/2}$

EXAMPLE OF GAS ORIFICE FLOW

Figure 10. System Diagram

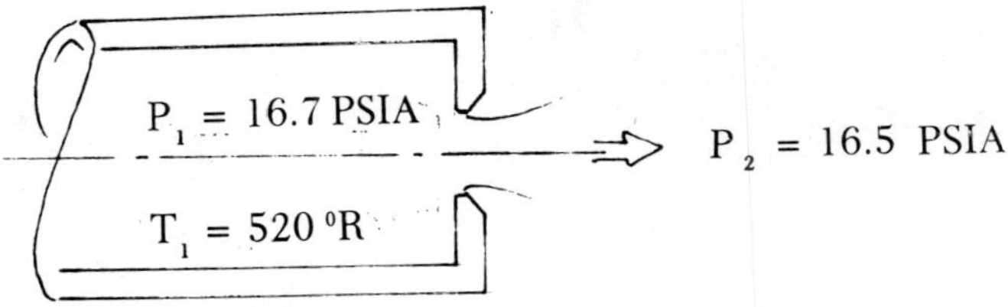

Area = 0.10 in^2
C_d = 0.95
Ratio of Specific Heats, k = 1.4

Figure 11.1 Calculation Results, Constant Downstream Pressure

PSIA	PSIA	LB/SEC
17.	16.7	0.0160
18.		0.0241
20.		0.0330
25.		0.0491
30		0.0633
40		0.0883

Figure 11.2 Calculation Results, Constant Upstream Pressure

P_1 PSIA	P_2 PSIA	dW LB/SEC
16.7	16.5	0.0143
	15.	0.0244
	14.	0.0274
	12.	0.0315
	10.	0.0344
	8.	0.0369
	4.	0.0369
	2.	0.0369

Figure 12.1 Analysis Plot, Constant Downstream Pressure

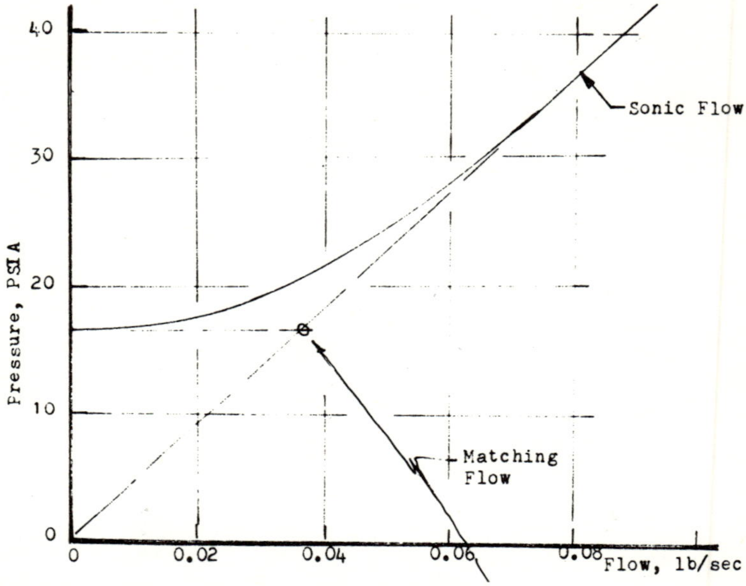

Figure 12.2 Analysis Plot, Constant Upstream Pressure

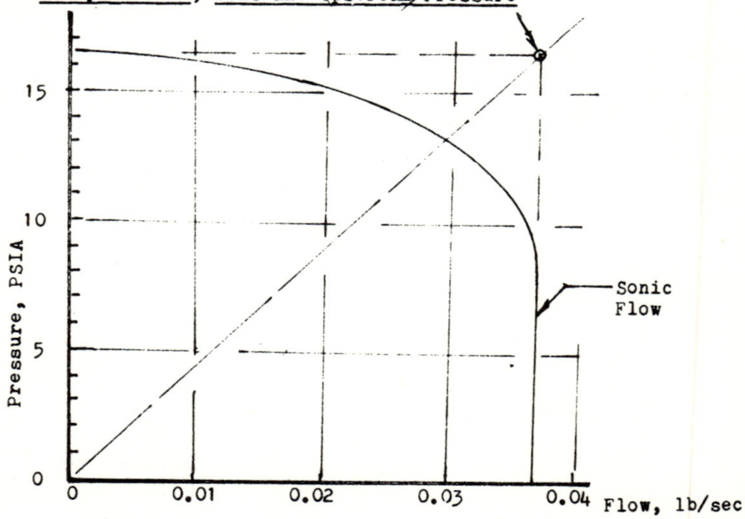

Figure 13. System Diagram
Air System
Viscosity = 3.7×10^{-7} lb-sec/ft²

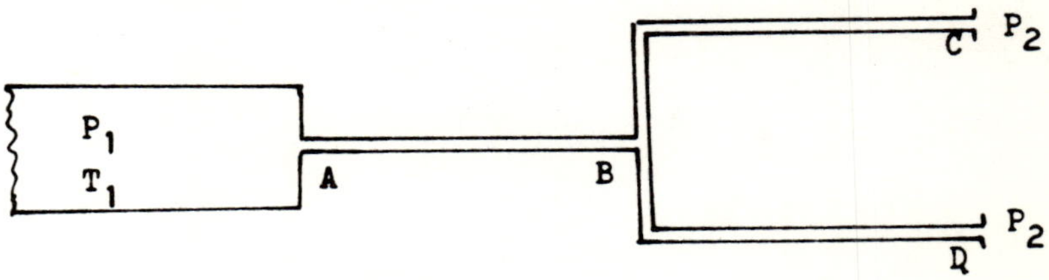

Location	Length ft	Dia ft	K	Area ft²	in²
A-B	10.0	0.3568	4.0	0.10	14.4
B-C	15.0	0.2523	3.0	0.05	7.2
B-D	20.0	0.2985	2.0	0.07	10.08

Figure 14. Calculation Results

Location	P_1 PSIA	P_2 PSIA	Assumed C_d	dW lb/sec	Calc. C_d
A-B	20.0	14.7	0.60	3.377	— —
		14.7	— —	3.377	0.48
		14.7	0.48	2.70	— —
		16.0	0.48	2.52	— —
		18.0	0.48	2.12	— —
B-C	20.0	14.7	0.60	1.69	— —
	20.0		— —	1.69	0.52
	20.0		0.52	1.46	— —
	18.0		0.52	1.20	— —
	16.0		0.52	0.87	— —
B-D	20.0	14.7	0.60	2.36	— —
	20.0		— —	2.36	0.60
	20.0		0.60	2.36	— —
	18.0		0.60	1.94	— —
	16.0		0.60	1.41	— —

Figure 15. Analysis Plot

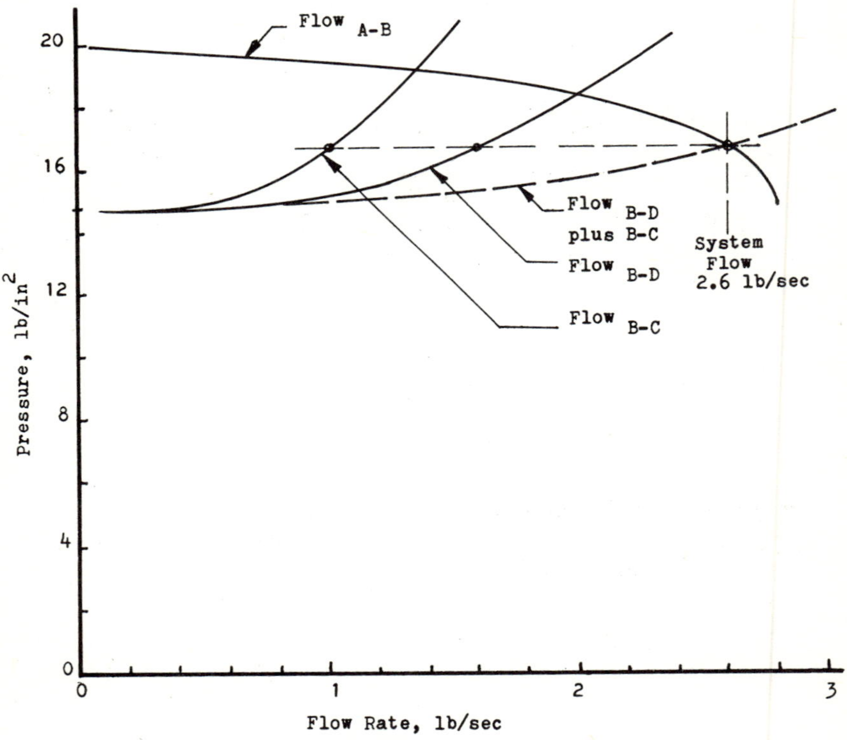

3.3 TWO-PHASE FLOW

Liquids commonly contain dissolved gas in the region between the molecules. In addition, liquids may contain "free" or undissolved gas.

Dissolved gas can be present in proportion to the partial pressure of the gas (the difference between the total pressure at the liquid-gas interface and the gas pressure.) The quantity of gas dissolved at equilibrium is affected by the temperature of the liquid and the affinity of the liquid for the particular gas as determined experimentally.

Free gas can be present as a result of the release of dissolved gas by a change of conditions which reduce gas solubility as well as by the entrainment of gas stirred in from the liquid-gas interface.

Bubbles of free gas produced by release of dissolved gas contain both vapor of the liquid and the dissolved gas in proportion to the partial pressures and the molecular weights. Bubbles are produced by heating the liquid or by lowering the static pressure.

Bubbles within a liquid rise to the surface at a rate determined by the difference in specific weight of the liquid and the gas and the diameter of the buble and the viscosity of the liquid according to Stokes' Law:

$$V = (2/9)(a^2/\mu)(g\rho_g - g\rho_l)$$

where

V = Bubble velocity, ft/sec (RN less than 1)

a = Bubble radius, feet

$g\rho_g$ = Specific Weight of gas, lb/ft^3

$g\rho_l$ = Specific Weight of Liquid, lb/ft^3

Gas released from solution must enter an existing nucleating bubble, see Bankoff, Reference. These nucleating bubbles are not in the bulk liquid but are attached to crevices at the wall surface. Because of surface tension liquids cannot wet the crevice to the apex. Highly polished surfaces are less susceptible to boiling and cavitation.

Release of dissolved gas can occur in a flowing liquid if the line pressure is reduced below the supply tank liquid-gas interface pressure. The process is known as two-phase flow. A wide spectrum of two-phase flow can exist, from a cloudy liquid, through increasingly large bubbles which tend to agglomerate into slugs having a rounded nose, to the end state of a misty vapor. Line configuration and gravity orientation have large effects. Useful applications are probably limited to a vapor/liquid ratio of 1/2 or less. This small value can be obtained almost entirely by the release of dissolved gas in most cases, having little effect on the vapor pressure of compound liquids.

Elevation changes and temperature differences which affect vapor pressure can be converted to equivalent pressure changes for simplicity of calculations.

The literature generally deals with two-phase flow in terms of vapor/liquid ratio which has a possible range of 0 to infinity. It is more convenient mathematically to deal with liquid-volume-fraction which has a possible range of 1.0 to 0.

$$V/L = (1/(LVF)) - 1$$

PROCEDURE FOR TWO-PHASE FLOW

(See Appendix I, Program Number 6)

1. Using trial liquid volume fraction, find two-phase pressure drop, based on data of Reference 11:

$$dP_{tp} = dP_{liq} \ (LVF)^{-0.905}$$

2. Find weight of air evolved per pound of liquid

$$dW_{air} = (dP_{tp} + dP_{head}) \ S_g$$

where dP ≐ Elevation head change and equivalent head vapor pressure change, lb/ft^2 ($ft \times lb/ft^3$)

S_g = Solubility coefficient, (lb gas/lb liq)/(lb/ft^2)

3. Find trial value of air partial pressure at line exit

$$P_{aex} = P_{ain} - dP_{tp} - dP_{head}$$

where P_{aex} = Pressure of air at exit, lb/ft^2

P_{ain} = Pressure of air at inlet, lb/ft^2

4. Find volume of released gas (air + vapor)

$$Vol_{air} = dW_{air} \ RT/P_{aex}$$

where Vol_{air} = Volume of air, ft^3

R_{air} = Gas Constant, 53.3 for air

T = Temperature, deg Rankine

For petroleum fuels $R_{vapor} = {}^1/{}_3 \ R_{air}$

Vapor is evolved in proportion to the partial pressure and the molecular weight, therefore

$$dW_{vapor} = dW_{air} \ (P_{vapor} / P_{aex})(RT/3RT))$$

$$Vol_{vapor} = dW_{vapor} \ (RT/(3P_{vapor}))$$

by combining the two above equations

$$Vol_{vapor} = dW_{air} \ RT/(9P_{aex})$$

The total gas evolved is

$$Vol_{gas} = dW_{air} \ R_{air} \ T(1 + (1/9))/P_{aex}$$

5. Find Liquid Volume Fraction

$$LVF = Vol_{liq} / (Vol_{liq} + Vol_{gas})$$

6. Test for agreement with original LVF value and if necessary, repeat the process until convergence is obtained.

Figure 16. System Diagram

Solubility of gas = 9.26×10^{-8} lb. gas/lb fuel/(lb/ft²)
Temp = 520°R
specific weight = 48. lb/ft³

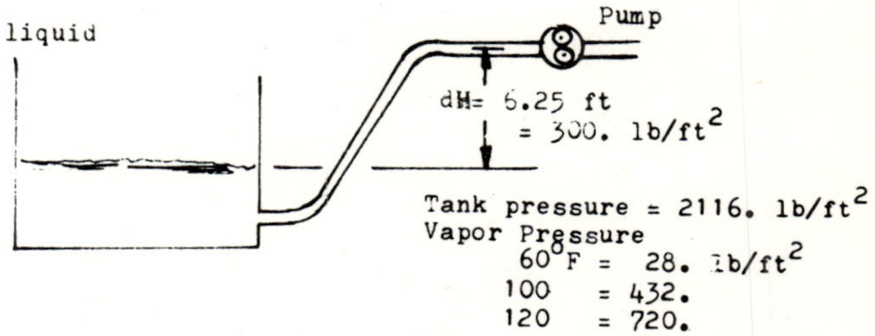

Figure 17. Calculation Results

°F	Temp °R	P_{air} PSFA	dP_{liq} PSFA	LVF	V/L	Relative Flow
60.	520.	2088.	10.	0.9651	0.04	0.11
			200.	0.9476	0.06	0.50
			400.	0.9302	0.08	0.71
			800.	0.8255	0.21	1.0
100.	560.	1684.	10.	0.9651	0.04	0.11
			200.	0.9302	0.08	0.50
			400.	0.8953	0.12	0.71
			800.	0.4415	1.27	1.0
120.	580.	1396.	10.	0.9476	0.06	0.11
			100.	0.9302	0.08	0.35
			200.	0.9127	0.10	0.50
			400.	0.8255	0.21	0.71

Figure 18. Analysis Plot

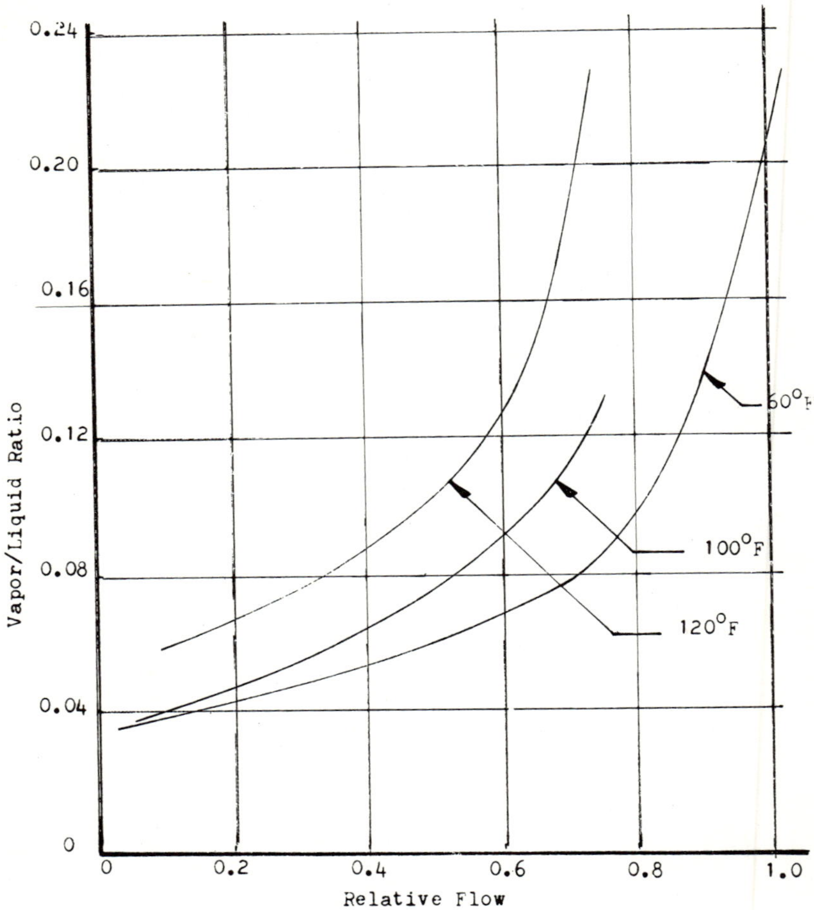

3.4 JET PUMP PERFORMANCE

Jet pump performance can be shown parametrically by a plot of pressure ratio and flow ratio:

$$N = \frac{P_d - P_o}{P_n - P_d}$$

where
N = Pressure Ratio
P_d = Discharge Pressure, lb/in^2
P_o = Secondary flow inlet pressure, lb/in^2
Pn = Primary nozzle pressure, lb/in^2

$$\emptyset = W_s / W_n$$

where
$\emptyset$ = Flow Ratio
W_s = Flow rate of secondary, lb/sec
W_n = Flow rate of nozzle, lb/sec

The performance of a jet pump is most strongly influenced by the ratio of the nozzle diameter to the mixing diameter. Important secondary factors are the shape of the secondary inlet, the distance of the nozzle from the inlet to the mixing section, the length of the mixing section, and the length of the exit diffuser.

The jet pump allows a high pressure fluid to escape through a nozzle giving it a high velocity pressure and a low static pressure. The low static pressure draws in the higher pressure secondary fluid which becomes frictionally entrained and mixed with the primary fluid. The fluids are decelerated in the diffuser with reduction of total energy.

$$\text{Eff.} = \emptyset N$$

where Eff. = Efficiency

PROCEDURE FOR JET PUMP PERFORMANCE
(See Appendix I, Program Numbers 7 and 8)

1. Find jet pump nominal pressure ratio at specified flow ratio:

$$N = \frac{2b + (1-2b)\dfrac{\phi^2 b^2}{(1-b)^2} - (1+k_{34})\, b^2 (1+\phi)^2}{(1+k_1) - \text{numerator}}$$

where b = Area ratio; (nozzle area)/(mixing area)

k_1 = Nozzle coefficient, use 0.2

k_{34} = Diffuser coefficient, use 0.35

N = Pressure ratio

ϕ = Flow ratio

2. Find discharge pressure for specified flow ratio

$$P_d = (N P_n + P_o)/(1+N)$$

where P_d = Discharge Pressure, lb/in^2

P_n = Nozzle Pressure, lb/in^2

P_o = Inlet Pressure, lb/in^2

3. Find nozzle flow rate

$$dQ_n = C_d A_n (2g H_n)^{1/2}$$

where dQ_n = Nozzle flow rate, ft^3/sec

C_d = Nozzle flow coefficient, use 0.95

A_n = Nozzle area, ft^2

H_n = Nozzle inlet total head, feet

4. Find secondary flow rate

$$dQ_s = \phi\, dQ_n$$

Figure 19. System Schematic

Nozzle Pressure = 10. PSI
Nozzle Diameter = 0.10 inch
Nozzle Flow = 2.27 x 10^{-3} ft^3/sec

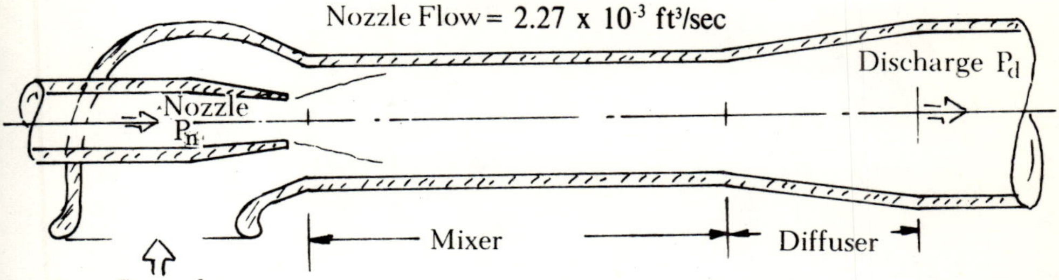

Figure 20. Calculation Results

Area Ratio b	Flow Ratio Ø	Pressure Ratio N	Secondary Flow ft^3/sec	Discharge P_d (PSI)
0.20	0.01	0.40	2.27×10^{-5}	2.86
	0.50	0.32	1.14×10^{-3}	2.42
	0.66	0.29	1.50	2.25
	1.16	0.20	2.63	1.67
	1.51	0.14	3.43	1.29
0.40	0.01	0.93	2.27×10^{-5}	4.82
	0.20	0.70	4.54×10^{-4}	4.12
	0.50	0.39	1.14×10^{-3}	2.81
	0.75	0.19	1.70	1.60
	1.00	0.02	2.27	0.20
0.90	0.01	1.30	2.27×10^{-5}	5.56
	0.03	0.94	6.81×10^{-5}	4.85
	0.05	0.56	1.14×10^{-5}	3.59
	0.06	0.39	1.36×10^{-4}	2.81
	0.08	0.10	1.82×10^{-4}	0.91

Figure 21.1 Analysis Plot, Pressure, Flow Ratios

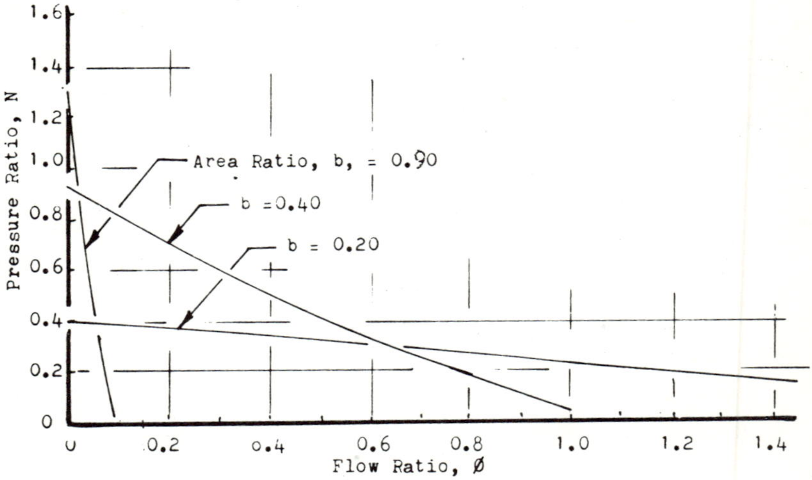

Figure 21.2 Analysis Plot, Operating Pressure and Flow

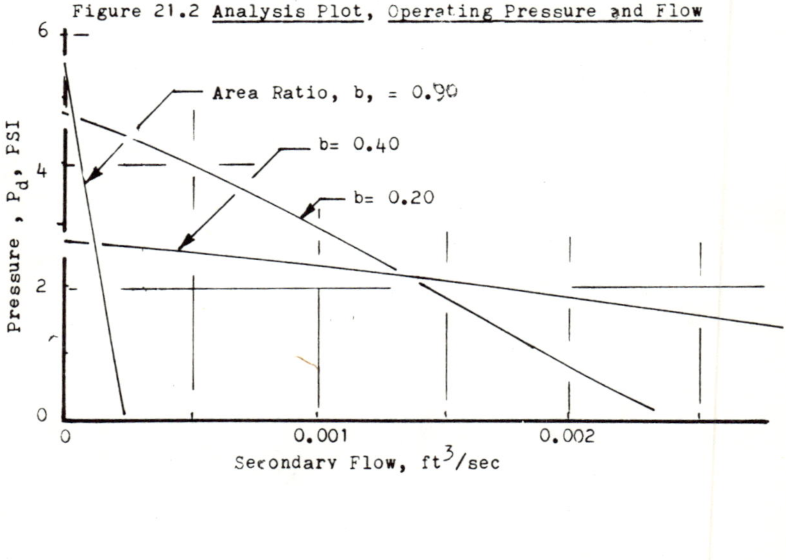

3.5 SURGE PRESSURE

Surge pressure is simply the conversion of liquid mass velocity energy into pressure energy. The conversion occurs at a rate governed by the sound velocity of the system.

Newton's law relating force and energy, F = MA, can be expressed as head and velocity:

$$\text{Force} = \text{Mass} \times \text{Acceleration}$$

$$(dH \times SpWt \times Area) = (Area \times Length \times SpWt/g) \times (dV/(Length/a))$$

$$dH = dV\,(a/g)$$

or

$$dH/dV = a/g$$

where dh = Change in pressure head, feet
dV = Change in mass velocity, feet/sec
a = Acoustic Velocity, feet/second
g = Gravity constant, 32.2 feet/sec ^{2}

The above expression is known as Joukowsky's Law after the developer, in Moscow, 1898. It forms the basis for the graphical analysis of surge. It shows that the change in pressure is directly proportional to the change in velocity. On a plot of mass velocity versus head a straight line of the slope a/g, with a plus or minus direction, can be used to connect supply and pressure loss curves at successive time intervals and thus describe a history of the surge process. The time intervals can be whole or fractional values of the time required for one pressure wave to leave and return to the valve.

$$\mu = 2L/a$$

where μ = Time for one acoustic cycle, seconds
L = Length of pipe from valve to tank, feet

The acoustic velocity is a composite found by a formula given later according to the compressibility of the liquid and the pipe material and the diameter/thickness ratio of the pipe.

Surge analysis can be separated into cases of "sudden" or "instant" opening or closing of a valve, or of "gradual" opening or closing of a valve. The sudden cases are applicable if the valve completely opens or closes in less than one acoustic cycle.

Sudden closing produces a pressure rise equal to the product of the original mass velocity times the ratio a/g minus the line friction loss.

Sudden opening results in accelerating the liquid at a rate set by the head difference between supply and flow pressure drop at the slope a/g. The mass velocity continues to increase until equilibrium is established between supply head and line flow head losses.

Gradual closing produces a succession of pressure waves until complete closure is obtained. After closure the surge pressures persist until damped by viscous forces as the mass velocity cyclically reverses.

Surge during opening of a line partly filled with liquid can be greater than the total supply head. This results by acceleration of the liquid through the partly filled line until reaching the control valve which causes sudden deceleration.

TABLE VIII. SURGE PROPERTIES
(Ref. 5)

Material	Bulk Modulus lb/ft²
Water	4.23×10^7
Gasoline	2.16×10^7
Crude Oil	3.38×10^7
Steel	4.23×10^9
Aluminum	1.44×10^9
Copper	2.45×10^9

PROCEDURE FOR ACOUSTIC VELOCITY AND SURGE CYCLE TIME
(See Appendix I, Program Number 9)

1. Find Acoustic Velocity

$$a = ((g/g\rho)/((D/(E_p t_p)) + (1/B_{liq})))^{1/2}$$

where
a = Acoustic Velocity, ft/sec
$g\rho$ = Liquid specific weight, lb/ft³
g = Gravity constant, 32.2 ft/sec²
D = Pipe Diameter, feet
E_p = Modulus of expansion of pipe, lb/ft²
B_{liq} = Bulk modulus of liquid, lb/ft²
t_p = Thickness of wall of pipe, feet

2. Find surge cycle time

$$\mu = 2L/a$$

where
μ = Cycle time for acoustic wave, seconds
L = Length from valve to tank, feet

EXAMPLE OF SURGE

Figure 22. System Diagram

Specific Weight = 48. lb/ft³
Pipe Diameter = 0.41 ft
Pipe Modulus = 1.44×10^9 lb/ft²
Liquid Modulus = 2.16×10^7 lb/ft²
Pipe Thickness = 4.08×10^{-3} ft

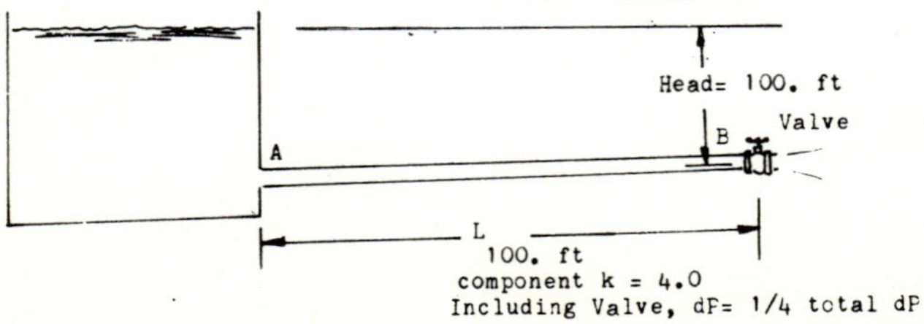

Figure 23. Calculation Results

Sound Velocity, a = 2404. ft/sec
Time of cycle, 2L/a = 0.0832 sec = μ
Rate of change of head, a/g = 74.66 sec

Flow Rate	Velocity	Line Pressure Drop			Max Surge	
lb/sec	ft/sec	PSF	PSI	Feet	Feet	PSI
50.	7.89	546	3.79	11.38	589.	196.
75.	11.84	1197.	8.36	24.94	885.	295.
100.	15.79	2089.	14.51	43.52	1179.	399.
200.	31.57	7996.	55.53	166.58	2357.	786.

EXAMPLE OF SURGE

Figure 24.1 Analysis Plot, Sudden Opening, Head vs. Velocity

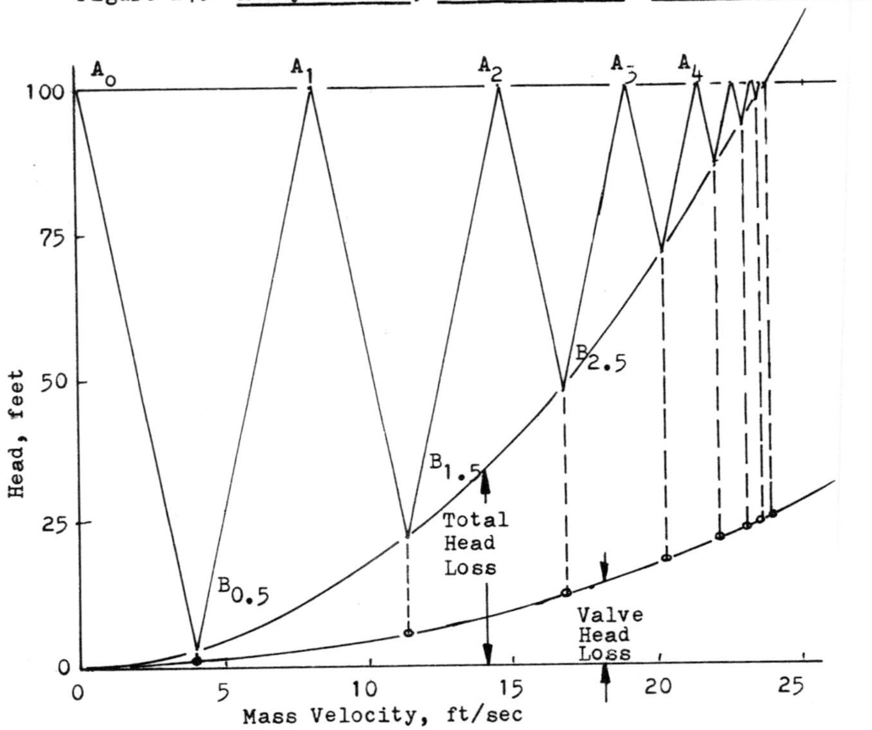

Figure 24.2 Analysis Plot, Sudden Opening, Head History

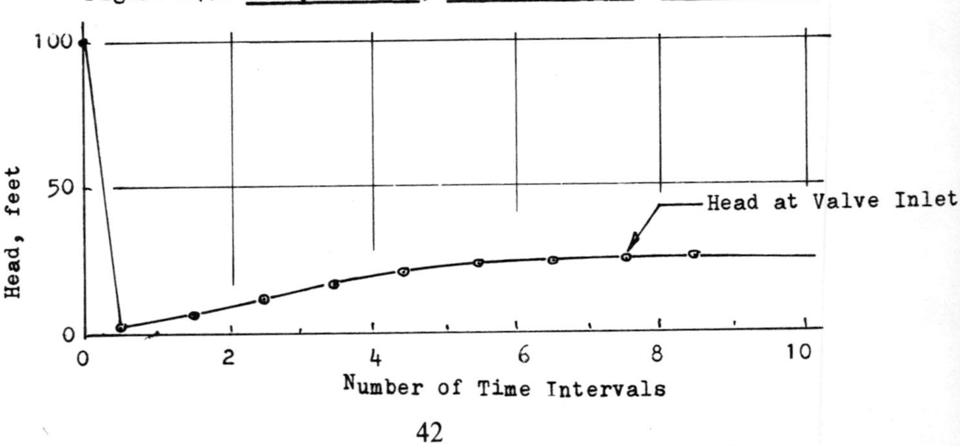

EXAMPLE OF SURGE

Figure 25.1 Analysis Plot, Gradual Opening, Head vs. Velocity

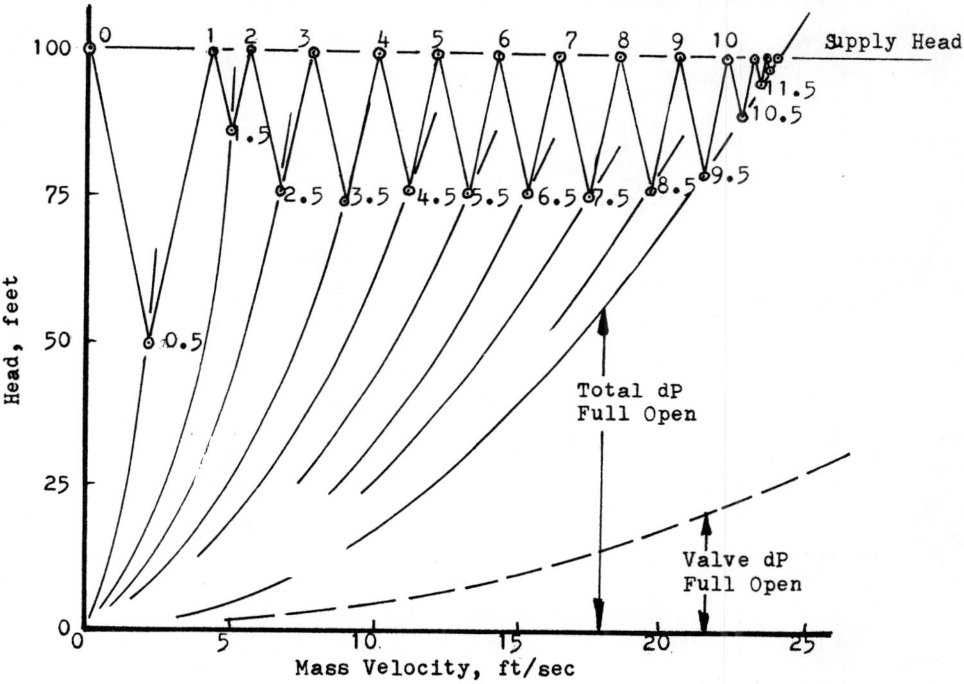

Figure 25.2 Analysis Plot, Gradual Opening, Head History

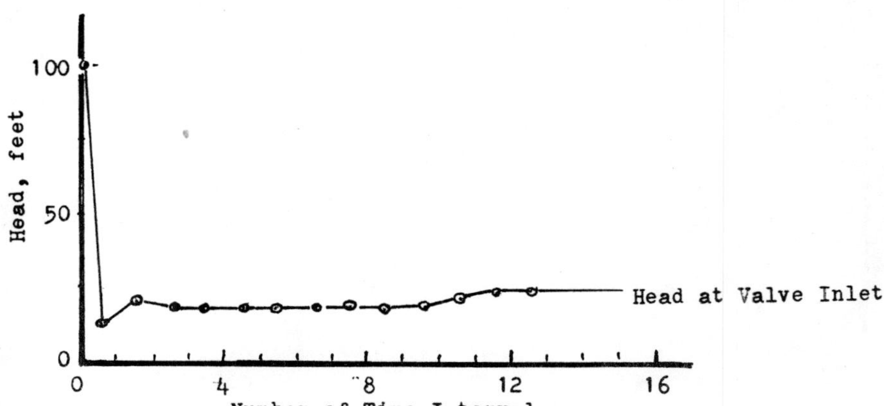

EXAMPLE OF SURGE

Figure 26.1 Analysis Plot, Sudden Closing, Head vs. Velocity

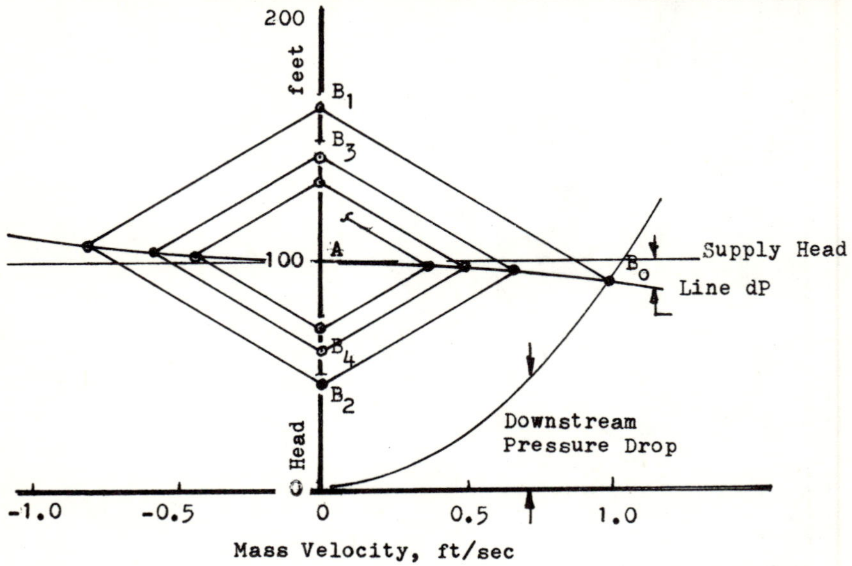

Figure 26.2 Analysis Plot, Sudden Closing, Head History

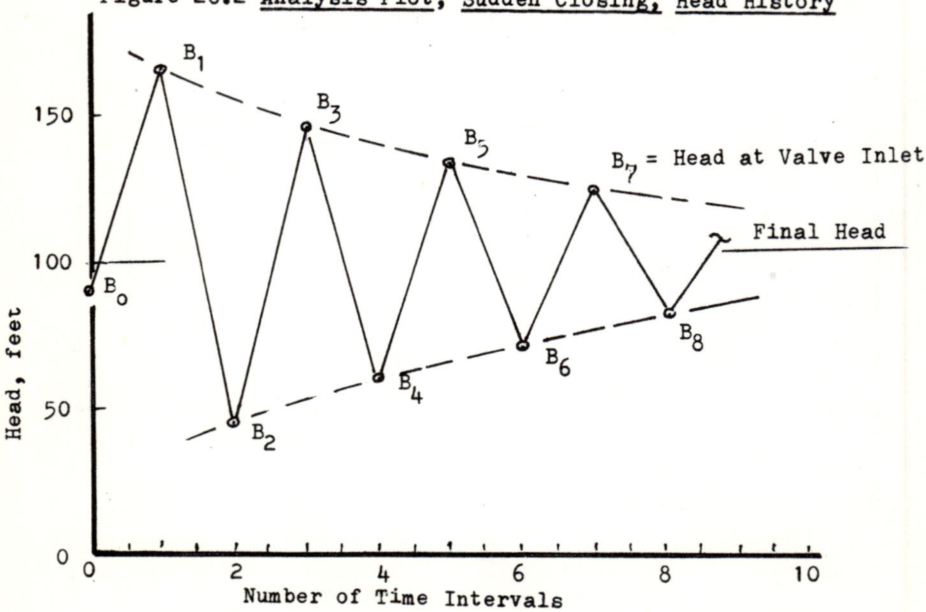

EXAMPLE OF SURGE

Figure 27.1 Analysis Plot, Gradual Closing, Head vs. Velocity

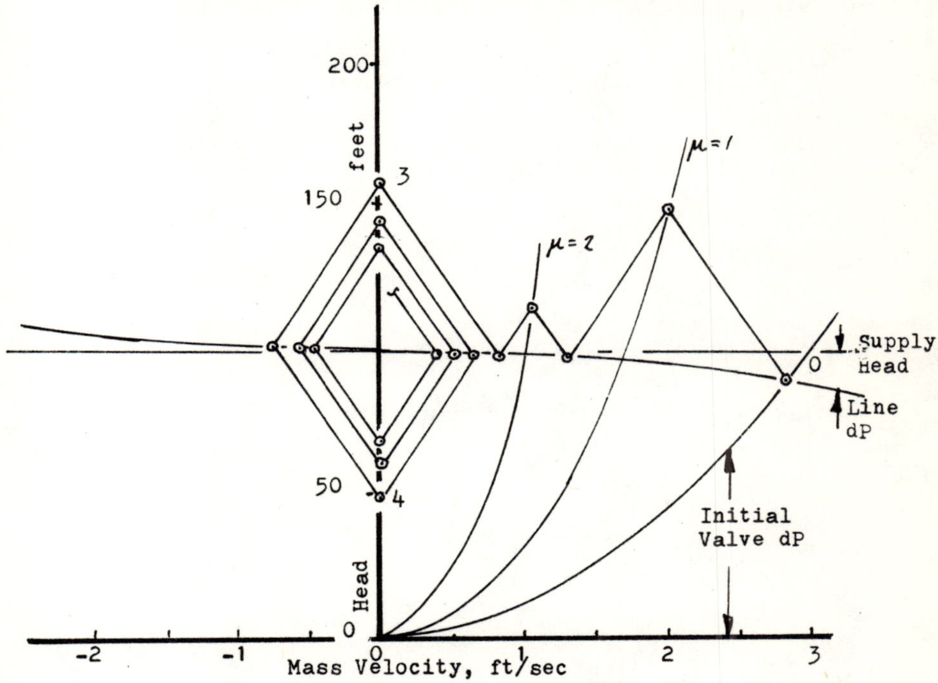

Figure 27.2 Analysis Plot, Gradual Closing, Head History

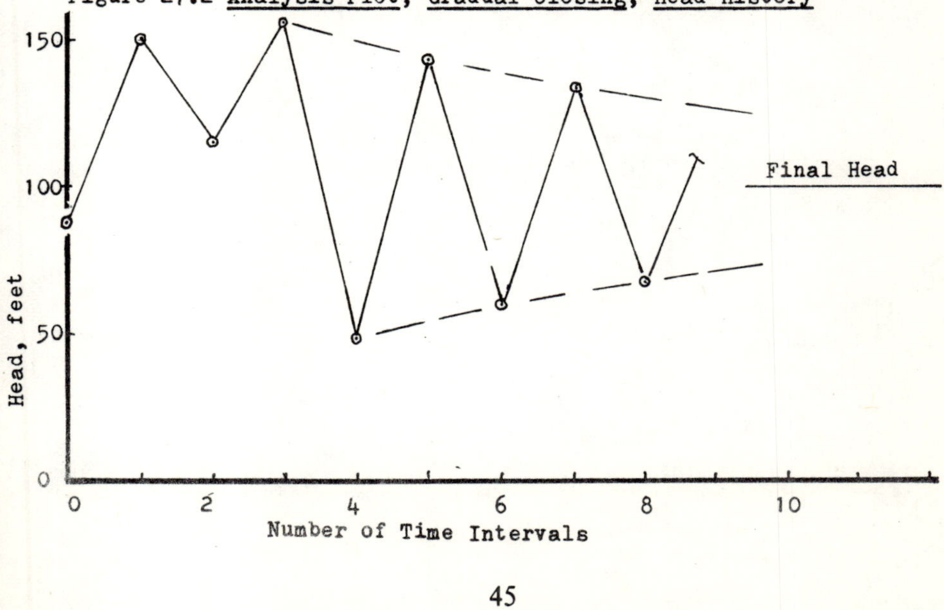

3.6 LIQUID PRESSURE REGULATOR FLOW

Pressure regulator designs are based upon springs and differential areas which establish the initial pressure at which flow occurs as well as the increase in flow rate with increasing pressure. There is a limiting flow rate above which the regulator is "wide open" and performs as a simple fixed orifice.

In devices which have a valve poppet attached to a circular diaphragm the effective diameter of the diaphragm is measured half the distance between the inner and outer fixed supports. The ratio of upstream to downstream areas plus any spring load determines the cracking pressure of the regulator (the initial flow pressure). For example, if the inlet pressure is 20 psi and the downstream/upstream area ratio is 3 then the cracking pressure is 60 psi (pounds per square inch).

PROCEDURE FOR LIQUID PRESSURE REGULATOR FLOW
(See Appendix I, Program Number 10)

Full open flow area must be known.
Initial cracking pressure must be known.
Head to obtain full open flow area must be known.

1. Test specified head to determine if regulator is full open, partly open, or closed.
2.. Full open flow rate

$$dQ = C_d A (8.03) H^{1/2}$$

where DQ = Flow rate, ft^3/sec
C_d = Discharge Coefficient, 0.6 to .9
A = Flow Area at full open, ft^2
$8.03 = (2g)^{1/2}$
H = Head difference, feet

3. Partly open flow rate

$$dQ=((H\text{-}H_o)/(H_c\text{-} H_o))\ dQ_o$$

where dQ_o = Flow rate at critical full open head, ft^3/sec
H_o = Head at cracking pressure, feet
H_c = Head at full open, feet

EXAMPLE OF LIQUID PRESSURE REGULATOR

Figure 28. System Diagram

Regulator
Full open flow area $= 0.0291\ ft^2$
Discharge coeff. C_d = 0.90
Head for full open = 19.24 feet
Flow at full open = 0.92 ft^3/sec
Head at cracking = 18.0 feet
Flow at part open = 0.74 $(ft^3/sec)/ft$

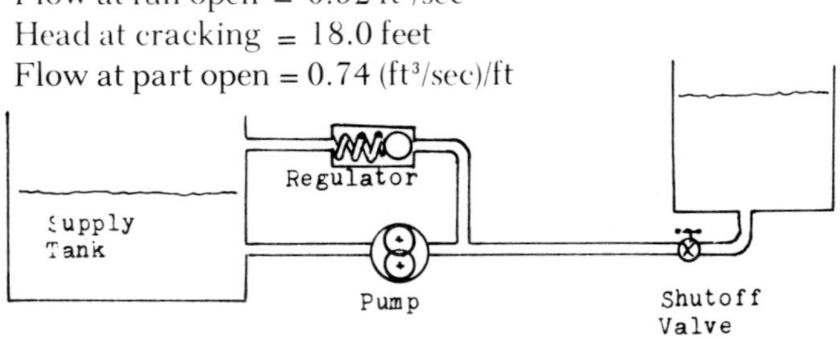

Figure 29. Calculation Results

Head feet	Flow Rate ft^3/sec
0	0
17.9	0
18.1	0.07
19.0	0.74
19.24	0.92
20.0	0.94
50.0	1.49
100.0	2.11
200.0	2.98

EXAMPLE OF LIQUID PRESSURE REGULATOR
Figure 30. Analysis Plot

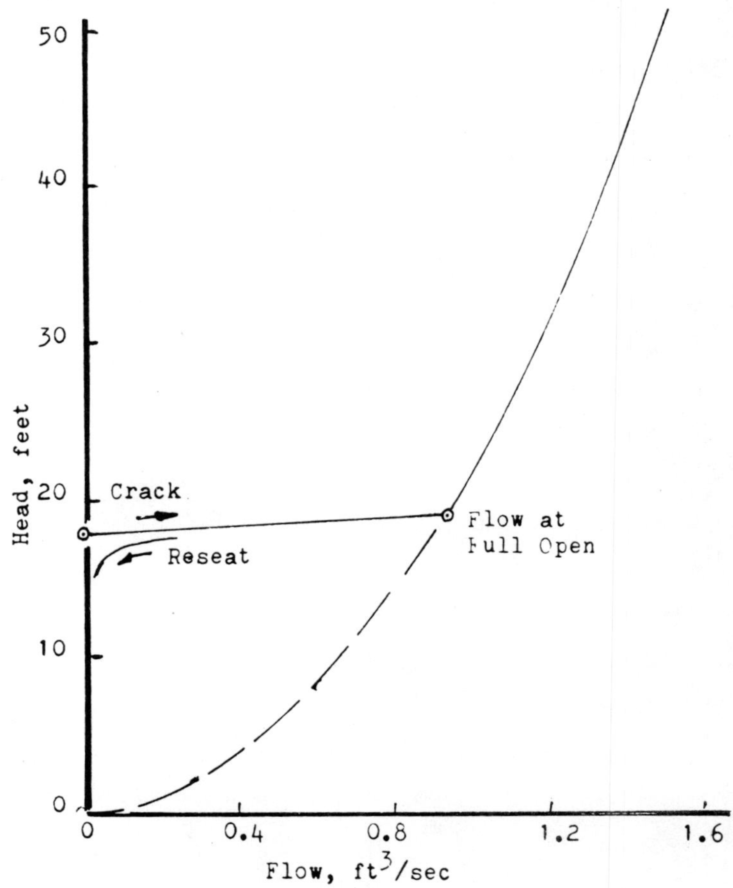

3.7 TANK EMPTYING INTERNAL FLOW

The case under consideration consists of a tank having a constant external emptying rate and one internal baffle which divides the tank into two compartments. The baffle has an orifice interconnector of known size and characteristics. The volume of each compartment and the volume/height factors are known.

A series of calculations are necessary to find the emptying history because the levels are constantly changing. Tests are necessary to avoid calculation results which might erroneously allow over-transfer or emptying of either compartment.

PROCEDURE FOR TANK EMPTYING INTERNAL FLOW

(See Appendix I, Program Number 11)

1. Find head difference at baffle

$$dH = Vol_1/(Vol/ft)_1 - Vol_2/(Vol/ft)_2$$

2. Test for negative head

If dH is negative, flow is zero at baffle

3. Find baffle flowrate

$$Q_1 = C_d\,A\,(8.03)\,(dH)^{1/2}$$

4. Adjust volumes for baffle flowrate

$$Vol_1 = Vol_1 - Q_1\,dt$$
$$Vol_2 = Vol_2 + Q_1\,dt$$

where dt = time interval, seconds

5. Test for $Vol_1 = 0$.

Set = 0 if necessary to avoid negative quantity

6. Find net flow to Vol_2

$$Vol_2 = Vol_2 - Q_2\,dt$$

7. Test for $Vol_2 = 0$.

Set = 0 if necessary to avoid negative quantity

8. Count each pass (usually 1 second per pass)
9. Test for both compartments empty

Stop if both are empty

10. Return to step 1 for next time increment

Figure 31. System Diagram

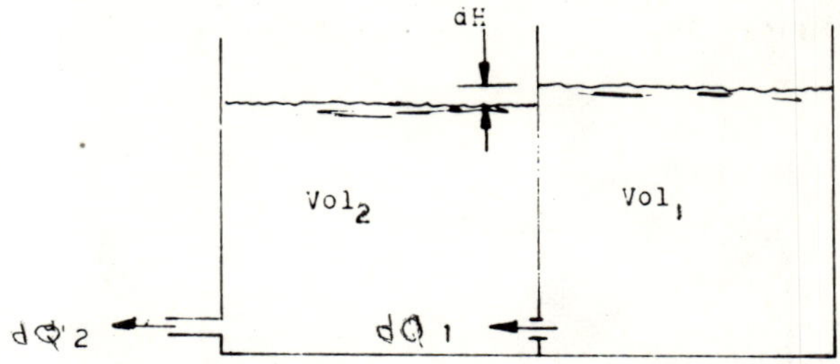

Flow Area$_1$ = 0.0404 ft^2

C_d = 0.65

dW_2 = 0.0926 ft^3/sec

Vol$_1$ = 3.0 ft^3 (Vol/ft)$_1$ = 19.0 ft^3/ft

Vol$_2$ = 2.9 ft^3 (Vol/ft)$_2$ = 19.0 ft^3/ft

Figure 32. Calculation Results

Interval	Time sec	Vol$_1$ ft^3	Vol$_2$ ft^3
0	0	3.00	2.90
1	1	2.94	2.77
7	7	2.79	2.46
23	23	2.23	1.54
36	36	1.69	0.88
48	48	1.16	0.30
56	56	0.75	0.04
64	64	0.42	0.03
79	79	0.08	0.00
89	89	0	0

Figure 33. Analysis Plot

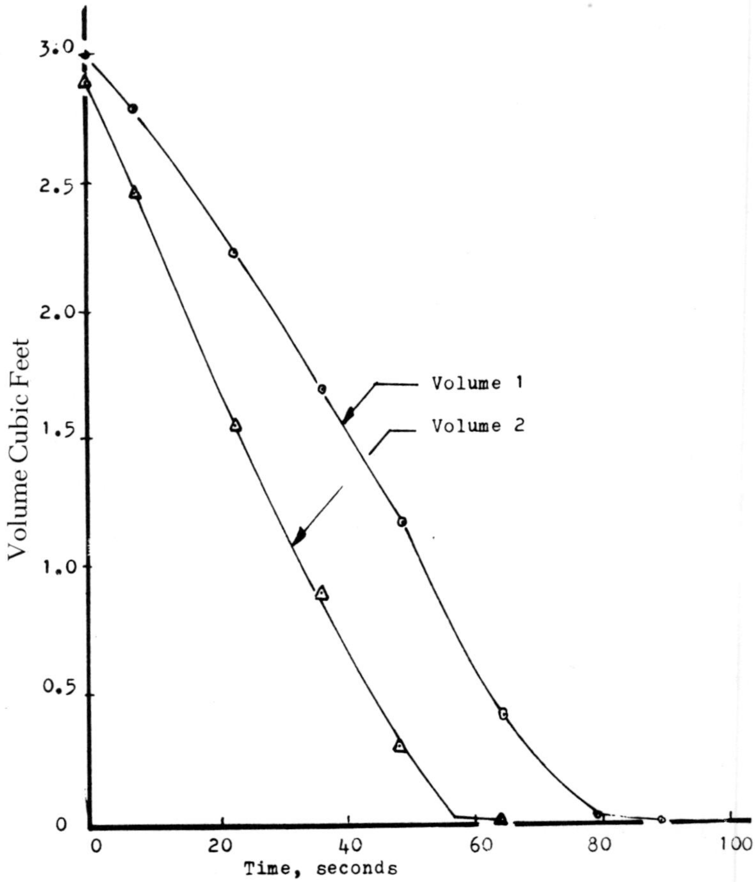

4.0 REFERENCES

1. "Centrifugal and Axial Flow Pumps", A. J. Stepanoff, John Wiley and Sons, New York, 1957
2. "Fluid Meters, Their Theory and Application", American Society of Mechanical Engineers, (ASME), 1959
3. "Jet Pump Theory and Performance With Fluids of High Viscosity", R. G. Cunningham, **Transactions of the ASME**, pp. 1807-1820, November, 1957
4. "Engineering Applications of Fluid Mechanics", J. C. Hunsaker and B. G. Rightmire, McGraw-Hill, New York, 1947
5. "Handbook of Engineering Fundamentals", O. W. Eshbach and Mott Souders, John Wiley & Sons, 1974
6. "Airplane Heating and Ventilating Equipment, Engineering Data", **SAE Aeronautical Information Report No. 23**, Society of Automotive Engineers, New York, 1951
7. "Pressure Losses in Smooth Pipe Bends", H. Ito, Journal of Basic Engineering, ASME, March, 1960, pp. 131-143
8. "Fluid Inlet Losses", C. W. Harris, **Bulletin No. 54**, University of Washington Experiment Station, October, 1930
9. "Engineering Data Sheets", Aero Supply Manufacturing Co., Corry, Pa.
10. "Ebullition From Solid Surfaces in the Absence of a Pre-Existing Gaseous Phase", S. G. Bankoff, **Transactions of the ASME**, pp. 735-740 May, 1957.
11. "A Pressure-Drop Correlation For Turbulent Two-Phase Flow of Gas-Liquid Mixtures in Horizontal Pipes", J. M. Chenoweth and M. W. Martin, ASME Paper No. 55-PET-9, 1955.
12. "HP-25 Owner's Handbook", Hewlet-Packard Company, Cupertino California, 1975
13. "Pressure-Density-Temperature Relations of Freezing Liquid ParaHydrogen to 350 Atmospheres", R. D. Goodwin and H. M. Roder, pp. 12-15, **Cryogenics**, Vol 3, No. 1, March, 1963
14. "Petroleum Refinery Engineering", W. L. Nelson, McGraw-Hill, 1958
15. "Thermodynamic Data on Oxygen and Nitrogen", J. Brewer, Air products and Chemicals, Inc., Allentown, Pa., ASD-TR-61-625, September 1961
16. "Jet Fuels", D. J. Keeling, **British Petroleum Aviation**, London, 1970
17. "Investigation of Solubility of Gases in Fuels", Thompson Products (TRW), Inglewood, Calif. IN-ER-183, 1955
18. "Physical Constants of Low Boiling Point Hydrocarbons and Miscellaneous Compounds"', R. Matteson and W. S. Hanna, Standard Oil Company of California, Presented at California Natural Gas Association Meeting, April, 1942

APPENDIX I, PROGRAMS FOR FLUID FLOW

Contents

DESCRIPTION OF CALCULATOR PROGRAM METHODOLOGY

Pocket calculators are manufactured according to two different mathematical concepts. One of these uses keys labelled with open and closed parenthesis to designate the order of computation. Equations are keyed in as written down. Ascending orders of parenthesis are needed for more complex equations. Intermediate results may not be visible.

The other concept, known as Reverse Polish Notation, RPN, dispenses with parentheses by keying in values and instructions to be performed consecutively but with the instruction following the value or values. The RPN concept is highly useful for calculators having small storage capacity

Additional instructions for a compatible machine are given in Reference 12.

PROGRAM ABBREVIATIONS AND INSTRUCTIONS

INPUT	Key in appropriate value
ENTER	Hold value for next operation
ROLL	Recall previous value in calculating registers
INTEGER	Delete fractional part
FRAC	Delete integral part
RCL	Recall from indicated location
STO	Store value in indicated location
STO +	Add to value in indicated location
STO —	Subtract from value in indicated location
STO X	Multiply by value in indicated location
STO DIVIDE	Divide value in indicated location
SIN	Sine of value
IF(X MORE THAN 0) (typical)	If "yes" go to next step; if "no" skip next step

HP-25 PROGRAM NUMBER 1

LIQUID PRESSURE DROP, TURBULENT FLOW, SMOOTH PIPES

Location	Name	Value	Dimension
0	$L.C_2$	100.184	feet.---
1	K	3.	-----
2	$C_1.\mu$	3000.00008	---.lb-sec/ft^2
3	g	32.2	ft/sec^2
4	C_3	-0.20	-----
5	$g\rho/g$	1.49	lb/ft^{3}32.2
6	Dia	0.25	feet
7	Velocity	-----	feet/sec

Step	Instruction	Stack	Note
00	INPUT dW		= (13.0)
01	π		
02	DIVIDE		
03	4		
04	X		
05	RCL 6		
06	SQUARE		
07	DIVIDE		
08	RCL 5		
09	DIVIDE		
10	RCL 3		
11	DIVIDE		= V (5.52)
12	STO 7		
13	RCL 6		
14	X		
15	RCL 5		
16	X		
17	RCL 2		
18	FRACTION		
19	DIVIDE	Y	= RN (25702)
20	ENTER		
21	RCL 2		
22	INTEGER	X	
23	IF (X MORE THAN Y)		Test for Laminar
24	STOP		

Step	Instruction	Stack	Note
25	ROLL	Y	
26	RCL 4	X	
27	Y^x		
28	RCL O		
29	FRAC		
30	X		= f (0.0241)
31	RCL 0		
32	INTEGER		
33	X		
34	RCL 6		
35	DIVIDE		
36	RCL 1		
37	+		= f(L/D) + K (12.66)
38	2		
39	DIVIDE		
40	RCL 5		
41	X		
42	RCL 7		
43	SQUARE		
44	X		= dP (287.33)
45	STOP		
46	RCL 7		
47	STOP		

HP-25 PROGRAM NUMBER 2

LIQUID PRESSURE DROP, LAMINAR FLOW

Location	Name	Value	Dimension
0	L	100.	feet
1	K	3.	-----
2	$C_1 \cdot \mu$	3000.00008	-----.lb-sec/ft^2
3	g	32.2	ft/sec^2
4	C_3	64.	-----
5	$g\rho$	1.49	lb/ft^3/32.2
6	Dia	0.25	ft
7	Velocity	-----	ft/sec

Step	Instruction		Comment
00	INPUT DW		(1.30)
01	π		
02	DIVIDE		
03	4		
04	X		
05	RCL 6		
06	SQUARE		
07	DIVIDE		
08	RCL 5		
09	DIVIDE		
10	RCL3		
11	DIVIDE		= V (0.55)
12	STO 7		
13	RCL 6		
14	X		
15	RCL 5		
16	X		
17	RCL 2		
18	FRACTION		
19	DIVIDE		= RN (2570)
20	ENTER	Y	
21	RCL 2		
22	INTEGER	X	
23	IF X LESS THAN Y		(Test for
24	STOP		Laminar)

Step	Instruction		Comment
25	ROLL	X	
26	RCL 4	Y	
27	INVERT X/Y		
28	DIVIDE		= f (0.0249)
29	RCL O		
30	X		
31	RCL 6		
32	DIVIDE		
33	RCL 1		
34	+		= f(L/D) + K (12.96)
35	2		
36	DIVIDE		
37	RCL 5		
38	X		
39	RCL 7		
40	SQUARE		
41	X		= dP (2.94)
42	STOP		
43	RCL 7		= V (0.55)
44	STOP		

HP-25 PROGRAM NUMBER 3

LIQUID PRESSURE DROP, TURBULENT FLOW, ROUGH PIPES

Location	Name	Value	Dimension
0	L.	100.	feet
1	K	3.	-----
2	$C_1 . \mu$	3000.00008	--- .lb-sec/ft²
3	g	32.2	ft/sec²
4	f	.037	-----
5	$g\rho/g$	1.49	lb/ft³/32.2
6	Dia	0.25	feet
7	Velocity	-----	feet/sec

Step	Instruction		Comment
00	INPUT dW		= (13.0)
01	π		
02	DIVIDE		
03	4		
04	X		
05	RCL 6		
06	SQUARE		
07	DIVIDE		
08	RCL 5		
09	DIVIDE		
10	RCL 3		
11	DIVIDE		= V
12	STO 7		(5.52)
13	RCL 6		
14	X		
15	RCL 5		
16	X		
17	RCL 2		
18	FRACTION		
19	DIVIDE		= RN (25702)
20	ENTER	Y	
21	RCL 2		
22	INTEGER	X	
23	IF (X MORE THAN Y)		Test for
24	STOP		Laminar
25	RCL 4		= f (0.037)
26	RCL 0		
27	INTEGER		
28	X		
29	RCL 6		
30	DIVIDE		
31	RCL 1		
32	+		= f (L/D) + K)
33	2		(17.80)
34	DIVIDE		
35	RCL 5		
36	X		
37	RCL 7		
38	SQUARE		
39	X		= dP (404.05)
40	STOP		
41	RCL 7		= V (5.52)
42	STOP		

HP-25 PROGRAM NUMBER 4
GAS ORIFICE FLOW

Location	Name	Value	Dimension
0	C_1	0.53	-----
1	C_2	0.25	-----
2	-----	-----	-----
3	P_1	16.7	PSIA
4	AREA	0.10	SQUARE INCHES
5	Cd	0.95	-----
6	T_1	520.	DEGREES RANKINE (460 + Deg F)
7	P_2/P_1	-----	-----

Step	Instruction		Note
00	INPUT P_2		= (16.5)
01	ENTER		
02	RCL 3		
03	DIVIDE		= P_2/P_1
04	STO 7	Y	(0.9880)
05	RCL O	X	
06	IF (X MORE THAN Y)		Test for
07	GO TO 18		Sonic Flow
08	1		
09	RCL 7		
10	—		
11	RCL 0		
12	DIVIDE	Y	
13	RCL 1	X	
14	Y^X		
15	RCL 0		
16	X		
17	GO TO 19		Subsonic ∅ = (0.2055)
18	RCL 0		Sonic ∅
19	RCL 5		
20	X		
21	RCL 3		
22	X		
23	RCL 6		
24	SQUARE ROOT		
25	DIVIDE		
26	RCL 4		
27	X		= dW (0.0143 lb/sec)
28	STOP		

HP-25 PROGRAM NUMBER 5

EQUIVALENT ORIFICE COEFFICIENT FOR LINE FLOW

Location	Name	Value	Dimension.
0	-----	-----	-----
1	C_1	25.25	English Units
2	Dia	0.20	feet
3	μ	3.5×10^{-7}	lb-sec/ft^2
4	Length	100.	feet
5	k	3.	-----
6	C_2	0.184	-----
7	C_3	-0.20	-----

Step	Instruction		
00	INPUT dW		= (1.0)
01	RCL 1		
02	DIVIDE		
03	RCL 2		
04	DIVIDE		
05	RCL 3		
06	DIVIDE	Y	= RN (5.66×10^5)
07	RCL 7	X	
08	Y^X		
09	RCL 6		
10	X		= f (0.0130)
11	RCL 4		
12	X		
13	STO 0		
14	RCL 2		
15	RCL 5		
16	X		
17	STO + 0		
18	RCL 2		
19	RCL 0		
20	DIVIDE		
21	SQUARE ROOT		= C_d (0.324)
22	STOP		

Appendix I.

HP-25 PROGRAM NUMBER 6
TWO-PHASE FLOW

Location	Name	Value	Dimension
0	Sol. Gas	9.26×10^{-8}	lb. gas/lb fuel/(lb/ft^2)
1	$P_{air} \cdot Vol_{liq}$	1828.0208	lb/ft^2.ft^3/lb
2	RT.C_1	27716.905	(53.3 x Deg R). -----
3	LVF	1.0	-----
4	dH	300	lb/ft^2
5	dP_{liq}	432.	lb/ft^2
6	scratch	-----	-----
7	scratch	-----	-----

Step	Key	Note
00	-----	
01	RCL 3	Y
02	RCL 2	
03	FRAC	
04	CHG SIGN	X
05	Y^X	
06	RCL 5	
07	X	$= dP_{tp}$
08	STO 6	(432.)
09	RCL 4	
10	+	
11	RCL 0	
12	X	
13	STO 7	
14	RCL 1	
15	INTEGER	
16	RCL 6	
17	—	
18	RCL 4	
19	—	
20	STO 6	
21	1	
22	ENTER	
23	9	
24	DIVIDE	

Step	Key	Note
25	1	
26	+	
27	RCL 7	
28	X	
29	RCL 2	
30	INTEGER	
31	X	
32	RCL 6	
33	DIVIDE	$= Vol_{aid} + Vol_{vapor}$
34	STO 7	(1.90×10^{-3})
35	RCL 1	
36	FRAC	Y
37	STO 6	
38	+	
39	RCL 6	X
40	INVERT X/Y	
41	DIVIDE	Y = LVF (0.92)
42	RCL 3	X
43	IF(X LESS THAN Y)	Test for Convergence
44	STOP	
45	1	
46	SIN	
47	STO — 3	
48	GO TO 01	

HP-25 PROGRAM NUMBER 7

JET PUMP PRESSURE RATIO FOR LIQUID FLOW

Location	Name	Value	Dimension
0	b	0.2	-----(area ratio)
1	Ø	0.5	-----(flow ratio)
2	$(1 + k_1)$	1.2	-----(1 + nozzle coefficient)
3	$(1 + k_{34})$	1.35	-----(1 + throat,diffuser coefficient)
4	-----		
5	-----		
6	-----		
7	-----		

Step	Instruction	
00	-----	
01	RCL 1	
02	RCL 0	
03	X	
04	SQUARE	
05	STO 6	
06	1	
07	RCL 0	
08	—	
09	SQUARE	
10	STO DIVIDE 6	
11	2	
12	RCL 0	
13	X	
14	1	
15	—	
16	CHG SIGN	
17	STO X 6	
18	1	
19	RCL 1	
20	+	
21	SQUARE	
22	STO 7	
23	RCL 0	
24	SQUARE	
25	STO X 7	
26	RCL 3	
27	STO X 7	
28	RCL 7	
29	STO — 6	
30	2	
31	RCL 0	
32	X	
33	STO + 6	
34	RCL 2	
35	RCL 6	
36	—	Y
37	RCL 6	X
38	INVERT X/Y	
39	DIVIDE	= N (0.32,0.29)
40	STOP	
41	IF(X LESS THAN O)	Test for N = O
42	STOP	
43	2	
44	DIVIDE	
45	STO + 1	
46	RCL 1	= new Ø (0.66)
47	STOP	
48	GO T 01	

HP-25 PROGRAM NUMBER 8

JET PUMP PERFORMANCE FOR LIQUID FLOW

Location	Name	Value	Dimension
0	N	0.39	-----(pressure ratio)
1	P_n	10.0	Nozzle Pressure, PSI
2	P_o	0.0	Inlet Pressure, PSI
3	scratch	-----	

00	-----	
01	RCL 0	
02	RCL 1	
03	X	
04	RCL 2	
05	+	
06	STO 3	
07	1	
08	RCL 0	Y
09	+	
10	RCL 3	X
11	INVERT X/Y	
12	DIVIDE	$= P_d$ (2.81)
13	STOP	

HP-25 PROGRAM NUMBER 9

ACOUSTIC VELOCITY AND SURGE CYCLE TIME

Location	Name	Value	Dimension
0	SpWt	48.0	lb/ft^3
1	LiqMod	2.16×10^7	lb/ft^2
2	PipeMod	1.44×10^9	lb/ft^2
3	Pipe thickness	4.08×10^{-3}	ft
4	Dia	0.41	ft
5	g	32.2	ft/sec^2
6	Length	100.0	feet

Step	Instruction	
00	-----	
01	RCL 1	X
02	1/X	
03	ENTER	
04	RCL 4	
05	RCL 2	
06	DIVIDE	
07	RCL 3	
08	DIVIDE	
09	+	X
10	1/X	
11	RCL 5	
12	X	
13	RCL 0	
14	DIVIDE	
15	SQUARE ROOT	
16	STOP	= Acoustic Velocity (2403.9 ft/sec)
17	RCL 6	
18	INVERT	
19	DIVIDE	
20	2	
21	X	= Cycle Time (0.0832 sec)
22	STOP	

HP-25 PROGRAM NUMBER 10

LIQUID PRESSURE REGULATOR FLOW

Location	Name	Value	Dimension
0	$Head_o$	18.0	feet
1	Spring Rate	0.74	$(ft^3/sec)/ft$
2	Area x C_d x 8.03	0.2106	ft^2 x $(ft/sec)^{1/2}$ (Area = 0.0291 ft^2)
3	$Head_{crit}$	19.24	ft (C_d = 0.90)
4	dQ_{crit}	0.92	ft^3/sec
5	Head	100.0	feet
6	scratch	-----	

Step	Instruction	Notes
00	-----	
01	RCL 0	Y
02	RCL 5	X
03	IF(X MORE THAN Y)	Test for No Flow
04	GO TO 06	
05	GO TO 28	
06	RCL 3	Y
07	RCL 5	X
08	IF(X LESS THAN Y)	Test for Orifice Flow
09	GO TO 15	
10	RCL 5	
11	SQUARE ROOT	
12	RCL 2	
13	X	= 2.11 ft^3/sec
14	STOP	
15	RCL 5	
16	RCL 0	
17	—	
18	STO 6	
19	RCL 3	
20	RCL 0	
21	—	
22	RCL 6	
23	INVERT	
24	DIVIDE	
25	RCL 4	
26	X	
27	STOP	= Part Open Flow
28	0	
29	STOP	

HP-25 PROGRAM NUMBER 11
TANK EMPTYING INTERNAL FLOW

Location	Name	Value	Dimension
0	scratch	-----	
1	Count	0	----- (Seconds)
2	Q_2	0.0926	ft^3/sec (External Flow)
3	Vol_1	3.0	ft^3
4	Vol_2	2.9	ft^3
5	$(Vol/ft)_1$	19.0	ft^3/ft
6	$(Vol/ft)_2$	19.0	ft^3/ft
7	C_d x 8.03	0.2106	Area in ft^2

Step	Instruction	X	Note
00	-----		
01	RCL 3		
02	RCL 5		
03	DIVIDE		
04	RCL 4		
05	RCL 6		
06	DIVIDE		
07	—	X	= dH
08	IF(X MORE THAN 0)		(0.0053)
09	GO TO 11		
10	0		
11	SQUARE ROOT		
12	RCL 7		
13	X		$= Q_1$
14	STO— 3		(0.0153)
15	STO + 4		
16	STO 0		
17	RCL 3	X	$= Vol_1$ (29847)
18	IF(X MORE THAN 0)		Test for $Vol_1 = 0$
19	GO TO 22		
20	0		
21	STO 3		
22	RCL 4		
23	RCL 2		
24	—	X	
25	IF(X MORE THAN 0)		Test
26	GO TO 30		for
27	0		$Vol_2 = 0$
28	STO 4		
29	GO TO 32		
30	RCL 2		
31	STO -4		
32	1		
33	STO + 1		Count time
34	RCL 3		(2.9847)
35	RCL 4		(2.8227)
36	+	X	
37	IF(X = 0)		Both Empty
38	STOP		
39	GO TO 01		

SR-56 PROGRAM NUMBER 1

LIQUID PRESSURE DROP, TURBULENT FLOW, SMOOTH PIPES

Location	Name	Value	Dimension
0			
1	$L.C_1$	100.184	feet.---
2	k	3.	---
3	μ	0.00008	$lb\text{-}sec/ft^2$
4	g	32.2	ft/sec^2
5	C_2	-0.20	---
6	$g\rho/g$	1.49	$(lb/ft^3)/32.2$
7	Dia	0.25	feet
8	Velocity	-----	ft/sec
9	Rn	-----	---
t	Transition	3000.	---

SR-56 PROGRAM NUMBER 2

LIQUID PRESSURE DROP, LAMINAR FLOW

Location	Name	Value	Dimension
0			
1	L	100.	feet
2	k	3.	---
3	μ	0.00008	$lb\text{-}sec/ft^2$
4	g	32.2	ft/sec^2
5	C_1	64.	---
6	$g\rho/g$	1.49	$(lb/ft^3)/32.2$
7	Dia	0.25	feet
8	scratch		
9			
t	transition	3000.	---

SR-56 PROGRAM NUMBER 1

LIQUID PRESSURE DROP, TURBULENT FLOW, SMOOTH PIPES

ENTER DW = 13.0 lb/sec, RST, R/S

Step	Key	Result
	2nd, CP	
	LRN	
01	X	
02	4	
03	DIVIDE	
04	(	
05	2nd, π	
06	X	
07	(	
08	RCL	
09	7	
10	X^2	
11	)	
12	X	
13	RCL	
14	6	
15	X	
16	RCL	
17	4	
18	)	
19	EQUALS	= V 5.52 ft/sec
20	STO	
21	8	
22	X	
23	(	
24	RCL	
25	7	
26	X	
27	RCL	
28	6	
29	)	
30	DIVIDE	
31	RCL	
32	3	
33	EQUAL	= Rn, 25702.
34	STO	
35	9	
36	INV	
37	2nd, X>t	
38	7	
39	8	
40	RCL	
41	9	
42	Y^X	
43	RCL	
44	5	
45	X	
46	(	
47	RCL	
48	1	
49	INV	
50	2nd, INT	
51	)	
52	EQUAL	= f 0.0241
53	X	
54	(	
55	RCL	
56	1	
57	2nd, INT	
58	)	
59	DIVIDE	
60	RCL	
61	7	
62	+	
63	RCL	
64	2	
65	EQUAL	= f(L/D) 12.66
66	DIVIDE	
67	2	
68	X	
69	RCL	
70	6	
71	X	
72	(	
73	RCL	
74	8	
75	X^2	
76	)	
77	EQUALS	= dP 287.33
78	R/S	
79	RST	
	LRN	

SR-56 PROGRAM NUMBER 2

LIQUID PRESSURE DROP, LAMINAR FLOW

ENTER DW = 1.3 lb /sec , RST, R/S

2nd, CP LRN		
01	X	
02	4	
03	DIVIDE	
04	(	
05	2nd, π	
06	X	
07	(	
08	RCL	
09	7	
10	X^2	
11	)	
12	X	
13	RCL	
14	6	
15	X	
16	RCL	
17	4	
18	)	
19	EQUALS	= V 0.55 ft/sec
20	STO	
21	8	
22	X	
23	(	
24	RCL	
25	7	
26	X	
27	RCL	
28	6	
29	)	
30	DIVIDE	

31	RCL	
32	3	
33	EQUAL	= Rn 2570.
34	STO	
35	9	
36	2nd, X>t	
37	6	
38	6	
39	RCL	
40	5	
41	DIVIDE	
42	RCL	
43	9	
44	X	
45	RCL	
46	1	
47	DIVIDE	
48	RCL	
49	7	
50	+	
51	RCL	
52	2	
53	EQUAL	= f(L/D) + k 12.96
54	DIVIDE	
55	2	
56	X	
57	RCL	
58	6	
59	X	
60	(	

61	RCL	
62	8	
63	X^2	
64	)	
65	EQUAL	= dP 2.94 psf
66	R/S	
67	RST	
	LRN	

SR-56 PROGRAM NUMBER 3

LIQUID PRESSURE DROP, TURBULENT FLOW, ROUGH PIPES

Location	Name	Value	Dimension
0			
1	L	100.	feet
2	k	3.	---
3	μ	0.00008	$lb\text{-}sec/ft^2$
4	g	32.2	ft/sec^2
5	f	0.037	---
6	$g\rho/g$	1.49	$(lb/ft^3)/32.2$
7	Dia	0.25	feet
8	scratch		
9	scratch		
t	Transition	3000.	---

SR-56 PROGRAM NUMBER 4

GAS ORIFICE FLOW

Location	Name	Value	Dimension
0			
1	C_1	0.53	---
2	C_2	0.25	---
3	P_1	16.7	psia
4	Area	0.10	ft^2
5	C_d	0.95	---
6	T_1	520.	deg F + 460. = deg R
7	scratch		
8	scratch		
9	scratch		
t	transition	0.53	---

SR-56 PROGRAM NUMBER 3

LIQUID PRESSURE DROP, TURBULENT FLOW, ROUGH PIPES

ENTER DW = 13.0 lb/sec, RST, R/S

Step	Key	
	2nd, CP	
	LRN	
01	X	
02	4	
03	DIVIDE	
04	(	
05	2nd, π	
06	X	
07	(	
08	RCL	
09	7	
10	X^2	
11	)	
12	X	
13	RCL	
14	6	
15	X	
16	RCL	
17	4	
18	)	
19	EQUALS	= V 5.52 ft/sec
20	STO	
21	8	
22	X	
23	(	
24	RCL	
25	7	
26	X	
27	RCL	
28	6	
29	)	
30	DIVIDE	
31	RCL	
32	3	
33	EQUAL	= Rn 25702.
34	STO	
35	9	
36	INV	
37	2nd, X>t	
38	6	
39	4	
40	RCL	
41	5	
42	X	
43	RCL	
44	1	
45	DIVIDE	
46	RCL	
47	7	
48	+	
49	RCL	
50	2	
51	EQUAL	= f(L/D) + k 17.80
52	DIVIDE	
53	2	
54	X	
55	RCL	
56	6	
57	X	
58	(	
59	RCL	
60	8	
61	X^2	
62	)	
63	EQUAL	= dP 404. psf
64	R/S	
65	RST	
	LRN	

SR-56 PROGRAM NUMBER 4

GAS ORIFICE FLOW

ENTER P_2 = 16.5 psia, RST, R/S

ENTER P_2 = 4.0 psia, RST, R/S

2nd, CP	
LRN	

Step	Key	
01	DIVIDE	
02	RCL	
03	3	
04	EQUAL	$= P_2/P_1$ 0.988 Or 0.2395
05	STO	
06	7	
07	INV	
08	2nd, X>t	
09	3	
10	0	
11	RCL	
12	1	
13	X	
14	(	
15	(	
16	1	
17	-	
18	RCL	
19	7	
20	)	
21	DIVIDE	
22	RCL	
23	1	
24	)	
25	Y^X	
26	RCL	
27	2	
28	EQUAL	$= \phi$, Subson 0.2055
29	STO	
30	9	

Step	Key	
31	RCL	
32	5	
33	X	
34	RCL	
35	4	
36	X	
37	RCL	
38	3	
39	DIVIDE	
40	(	
41	RCL	
42	6	
43	SQRT	
44	)	
45	EQUAL	$= C_d \ A \ P/T^{1/2}$ 0.0695
46	STO	
47	8	
48	RCL	
49	7	
50	INV	
51	2nd, X>t	
52	6	
53	3	
54	RCL	
55	8	
56	X	
57	RCL	
58	9	
59	EQUAL	= dW, subsonic 0.0143 lb./sec
60	R/S	

Step	Key	
61	GO TO	
62	9	
63	9	
64	RCL	
65	1	
66	X	
67	RCL	
68	8	
69	EQUAL	= dW, sonic 0.0369 lb/sec
70	R/S	
	LRN	

SR-56 PROGRAM NUMBER 5

EQUIVALENT ORIFICE COEFFICIENT FOR LINE FLOW

Location	Name	Value	Dimension
0	scratch		
1	C_1	25.25	english units
2	Dia	0.20	ft
3		3.5×10^{-7}	$lb\text{-}sec/ft^2$
4	Length	100.	ft
5	k	3.	---
6	C_2	0.184	---
7	C_3	-0.20	---

ENTER dW = 1.0 lb/sec, RST, R/S

Step	Key	Result
	2nd, CP	
	LRN	
01	DIVIDE	
02	RCL	
03	1	
04	DIVIDE	
05	RCL	
06	2	
07	DIVIDE	
08	RCL	
09	3	
10	EQUAL	= Rn
11	Y^X	5.66×10^5
12	RCL	
13	7	
14	X	
15	RCL	
16	6	
17	EQUAL	= f, 0.013
18	X	
19	RCL	
20	4	
21	EQUAL	= fL, 1.3
22	STO	
23	0	
24	RCL	
25	2	
26	X	
27	RCL	
28	5	
29	EQUAL	= kD 0.6
30	SUM	
31	0	
32	RCL	
33	2	
34	DIVIDE	
35	RCL	
36	0	
37	EQUAL	
38	SQRT	
39	EQUAL	= C_d , 0.324
40	R/S	
	LRN	

SR-56 PROGRAM NUMBER 6

TWO-PHASE FLOW

Location	Name	Value	Dimension
0	Sol. Gas	9.26×10^{-8}	(lb gas/lb fuel)/(lb/ft^3)
1	P_{air}	1828.	lb/ft^2
2	Vol_{liq}	0.0208	ft^3/lb
3	R T	27716.	53.3 x deg R
4	C_3	-0.905	---
5	LVF	1.0	---
6	dH	300.	lb/ft^2
7	dP_{liq}	432.	lb/ft^2
8	scratch		
9	scratch		
t	test	1.0	

SR-56 PROGRAM NUMBER 7

JET PUMP PRESSURE RATIO AND LIQUID FLOW

Location	Name	Value	Dimension
0			
1	b	0.2	--- (Area Ratio)
2	$\dot{\phi}$	0.5	--- (Flow Ratio)
3	$(1 + k_1)$	1.2	--- (1 + Nozzle Coefficient)
4	$(1 + k_{3,4})$	1.35	--- (1 + Throat, Diffuser Coefficient)
5	---		
6	scratch		
7	scratch		
8			
9			
t	test	0.0	

SR -56 PROGRAM NUMBER 6

TWO-PHASE FLOW

2nd, CP		
LRN		
01	RCL	
02	5	
03	Y^X	
04	RCL	
05	4	
06	X	
07	RCL	
08	7	
09	EQUAL	= dP_{tp}
10	STO	432. psf
11	8	
12	RCL	
13	8	
14	+	
15	RCL	
16	6	
17	EQUAL	= 6.7×10^{-5}
18	X	
19	RCL	
20	0	
21	EQUAL	
22	STO	
23	9	
24	RCL	
25	1	
26	-	
27	RCL	
28	8	
29	-	
30	RCL	

31	6	
32	EQUAL	= 1096.
33	STO	
34	8	
35	1	
36	DIVIDE	
37	9	
38	+	
39	1	
40	EQUAL	
41	X	
42	RCL	
43	9	
44	X	
45	RCL	
46	3	
47	DIVIDE	
48	RCL	
49	8	
50	EQUAL	= 1.9×10^{-3}
51	STO	
52	9	
53	RCL	
54	9	
55	+	
56	RCL	
57	2	
58	EQUAL	
59	DIVIDE	
60	RCL	

61	2	
62	EQUAL	
63	2nd, 1/X	= LVF
64	2nd, X>t	0.916
65	7	
66	9	
67	X⇌t	
68	-	
69	.	
70	0	
71	1	
72	EQUAL	= 0.99
73	STO	
74	5	
75	X⇌t	
76	GO TO	
77	0	
78	0	
79	R/S	
	LRN	

SR-56 PROGRAM NUMBER 7

JET PUMP PRESSURE RATIO FOR LIQUID FLOW

Step	Key	Result
	2nd, CP	
	LRN	
01	RCL	
02	2	
03	X	
04	RCL	
05	1	
06	EQUAL	
07	SQUARE	= 0.01
08	STO	
09	6	
10	1	
11	-	
12	RCL	
13	1	
14	EQUAL	
15	SQUARE	= 0.64
16	INV	
17	2nd, PROD	
18	6	
19	2	
20	X	
21	RCL	
22	1	
23	EQUAL	
24	-	
25	1	
26	EQUAL	
27	+/-	= 0.60
28	2nd, PROD	
29	6	
30	1	
31	+	
32	RCL	
33	2	
34	EQUAL	
35	SQUARE	= 2.25
36	STO	
37	7	
38	RCL	
39	1	
40	SQUARE	
41	2nd, PROD	
42	7	
43	RCL	
44	4	
45	2nd, PROD	
46	7	
47	RCL	
48	7	
49	INV	
50	SUM	
51	6	
52	2	
53	X	
54	RCL	
55	1	
56	EQUAL	= 0.4
57	SUM	
58	6	
59	RCL	
60	3	
61	-	
62	RCL	
63	6	
64	EQUAL	= 0.912
65	DIVIDE	
66	RCL	
67	6	
68	EQUAL	
69	2nd, 1/x	= N 0.316
70	R/S	
71	INV	
72	2nd, x>t	
73	8	
74	6	
75	RCL	
76	2	
77	-	
78	.	
79	0	
80	5	
81	EQUAL	= 0.45
82	STO	
83	2	
84	GO TO	
85	0	
86	0	
87	R/S	
	LRN	

SR-56 PROGRAM NUMBER 8

JET PUMP PERFORMANCE FOR LIQUID FLOW

Location	Name	Value	Dimension
0	N	0.39	--- (Pressure Ratio)
1	P_N	10.0	Nozzle Pressure, psi
2	P_O	0.0	Inlet Pressure, psi
3	scratch		

2nd, CP		
LRN		
01	RCL	
02	0	
03	X	
04	RCL	
05	1	
06	EQUAL	
07	+	
08	RCL	
09	2	
10	EQUAL	$= (N\ P_N) + P_O$
11	STO	3.9
12	3	
13	1	
14	+	
15	RCL	
16	0	
17	EQUAL	= 1+N
18	DIVIDE	1.39
19	RCL	
20	3	
21	EQUAL	
22	2nd, 1/X	$= P_D$ 2.81 psi
23	R/S	
	LRN	

SR-56 PROGRAM NUMBER 9

ACOUSTIC VELOCITY AND SURGE CYCLE TIME

Location	Name	Value	Dimension
0	Specific Weight	48.0	lb/ft^3
1	Liq. Mod.	2.16×10^7	lb/ft^2
2	Pipe Mod.	1.44×10^9	lb/ft^2
3	Pipe thick.	4.08×10^{-3}	ft
4	Dia.	0.41	ft
5	g	32.2	ft/sec^2
6	Length	100.	ft
7	scratch		

2nd, CP

LRN

Step	Key	Result
01	RCL	
02	4	
03	DIVIDE	
04	(	
05	RCL	
06	2	
07	X	
08	RCL	
09	3	
10	)	
11	EQUAL	$= 6.98 \times 10^{-8}$
12	STO	
13	7	
14	RCL	
15	1	
16	2nd, 1/X	$= 4.63 \times 10^{-8}$
17	SUM	
18	7	
19	RCL	
20	5	
21	DIVIDE	
22	RCL	
23	0	
24	EQUAL	= 0.670
25	DIVIDE	
26	RCL	
27	7	
28	EQUAL	
29	2nd, SQRT	= Acoustic Velocity, 2404. ft/sec
30	R/S	
31	DIVIDE	
32	RCL	
33	6	
34	EQUAL	
35	2nd, 1/X	
36	X	
37	2	
38	EQUAL	= Cycle time, 0.0832 sec.
39	R/S	
	LRN	

SR-56 PROGRAM NUMBER 10

LIQUID PRESSURE REGULATOR FLOW

Location	Name	Value	Dimension	
0	$Head_0$	18.0	ft*	
1	Spring Rate	0.74	$(ft^3/sec)/ft$ of head	
2	(Area x C_d x 8.03)	0.2106	$ft^2 \times (ft/sec)^{1/2}$	Area= 0.0291 ft^2
3	$Head_{crit}$	19.24	ft	
4	dQ_{crit}	0.92	ft^3/sec	
5(case 1)	Head	100.	ft	
(case 2)		19.	ft	
6	scratch			
t	test			

2nd, CP

LRN

Step	Key	
01	RCL	
02	0	
03	X⇌t	
04	RCL	
05	5	
06	INV	
07	2nd, X>t	
08	5	
09	6	
10	RCL	
11	3	
12	X⇌t	
13	0	
14	RCL	
15	5	
16	2nd, X>t	
17	4	
18	5	
19	RCL	
20	5	
21	-	
22	RCL	
23	0	
24	EQUAL	
25	STO	
26	6	
27	RCL	
28	3	
29	-	
30	RCL	
31	0	
32	EQUAL	
33	DIVIDE	
34	RCL	
35	6	
36	EQUAL	= 1.24
37	2nd, 1/X	
38	X	
39	RCL	
40	4	
41	EQUAL	= Spring Flow 0.742 ft^3/sec
42	R/S	
43	GO TO	
44	0	
45	0	
46	RCL	
47	5	
48	2nd, SQRT	
49	X	
50	RCL	
51	2	
52	EQUAL	= Orifice Flow 2.106 ft^3/sec
53	R/S	
54	GO TO	
55	0	
56	0	
57	0	
58	R/S	
	LRN	

SR-56 PROGRAM NUMBER 11

TANK EMPTYING INTERNAL FLOW

Location	Name	Value	Dimension
0	scratch		
1	count	0	seconds
2	Q_2	0.0926	ft^3/sec (External Flow)
3	Vol_1	3.0	ft^3
4	Vol_2	2.9	ft^3
5	$(Vol/ft)_1$	19.0	ft^3/ft
6	$(Vol/ft)_2$	19.0	ft^3/ft
7	C_d x 8.03	0.2106	
8	scratch		
9	scratch		
t	test		

SR-56 PROGRAM NUMBER 11

TANK EMPTYING INTERNAL FLOW

Step	Key	Remarks
	2nd, CP	
	LRN	
01	RCL	
02	4	
03	DIVIDE	
04	RCL	
05	6	
06	EQUAL	= H_2 , 0.153 ft
07	STO	
08	8	
09	RCL	
10	3	
11	DIVIDE	
12	RCL	
13	5	
14	EQUAL	= H_1, 0.158 ft
15	-	
16	RCL	
17	8	
18	EQUAL	= dH, 0.0053 ft
19	STO	
20	8	
21	INV	
22	2nd, X>t	Test $dH < 0$
23	6	
24	4	
25	RCL	
26	8	
27	2nd, SQRT	
28	X	
29	RCL	
30	7	
31	EQUAL	= dQ_1 0.0153
32	STO	
33	0	
34	INV	
35	SUM	
36	3	
37	SUM	
38	4	
39	RCL	
40	3	= 2.9847
41	INV	
42	2nd, X>t	Test $Vol_1 < 0$
43	6	
44	8	
45	RCL	
46	4	= 2.915
47	-	
48	RCL	
49	2	
50	EQUAL	= 2.82
51	INV	
52	2nd, X>t	Test $Q_2 > Vol_2$
53	7	
54	4	
55	RCL	
56	2	
57	INV	
58	SUM	
59	4	
60	1	
61	SUM	
62	1	= End of Time Interval
63	R/S	
64	RST	
65	0	
66	GO TO	
67	3	
68	0	
69	0	
70	STO	
71	3	
72	GO TO	
73	4	
74	4	
75	0	
76	STO	
77	4	
78	RCL	
79	0	
80	STO	
81	2	
82	GO TO	
83	5	
84	9	
	LRN	

APPENDIX II. ADDITIONAL FLUID PROPERTIES

Figure

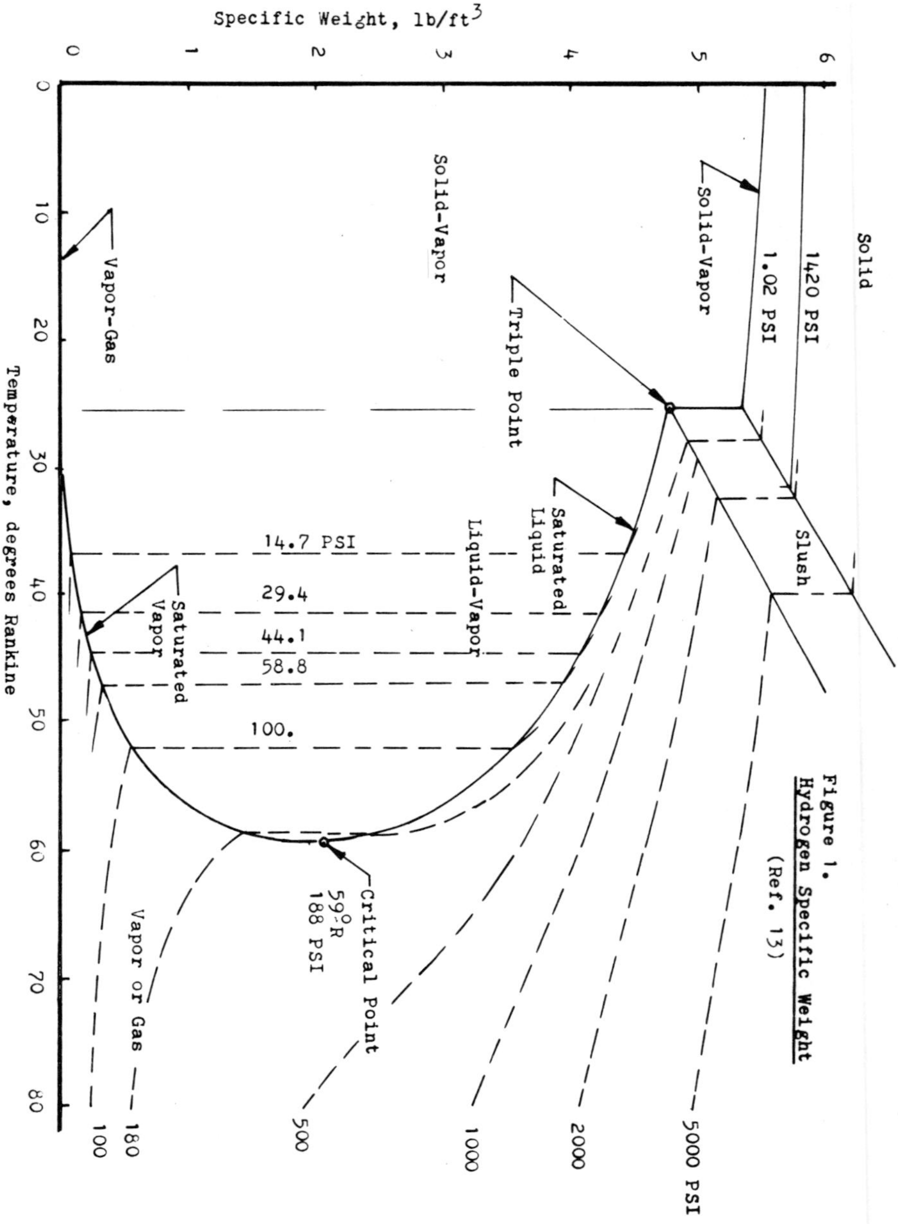

Figure 1.
Hydrogen Specific Weight
(Ref. 13)

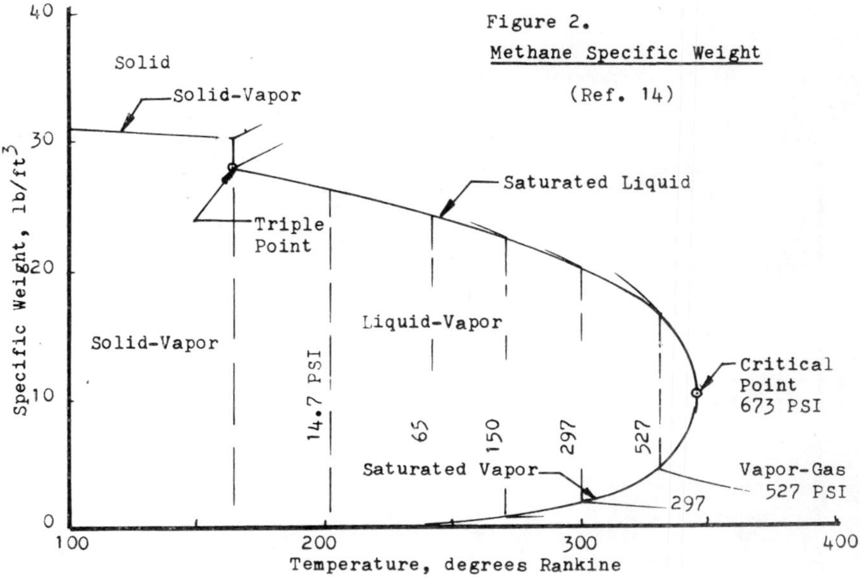

Figure 2.
Methane Specific Weight
(Ref. 14)

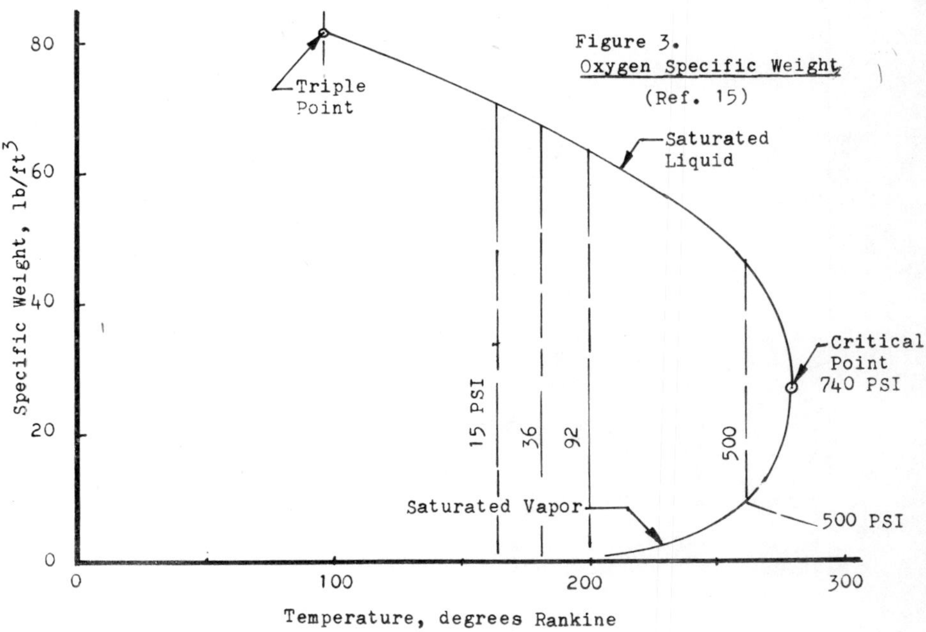

Figure 3.
Oxygen Specific Weight
(Ref. 15)

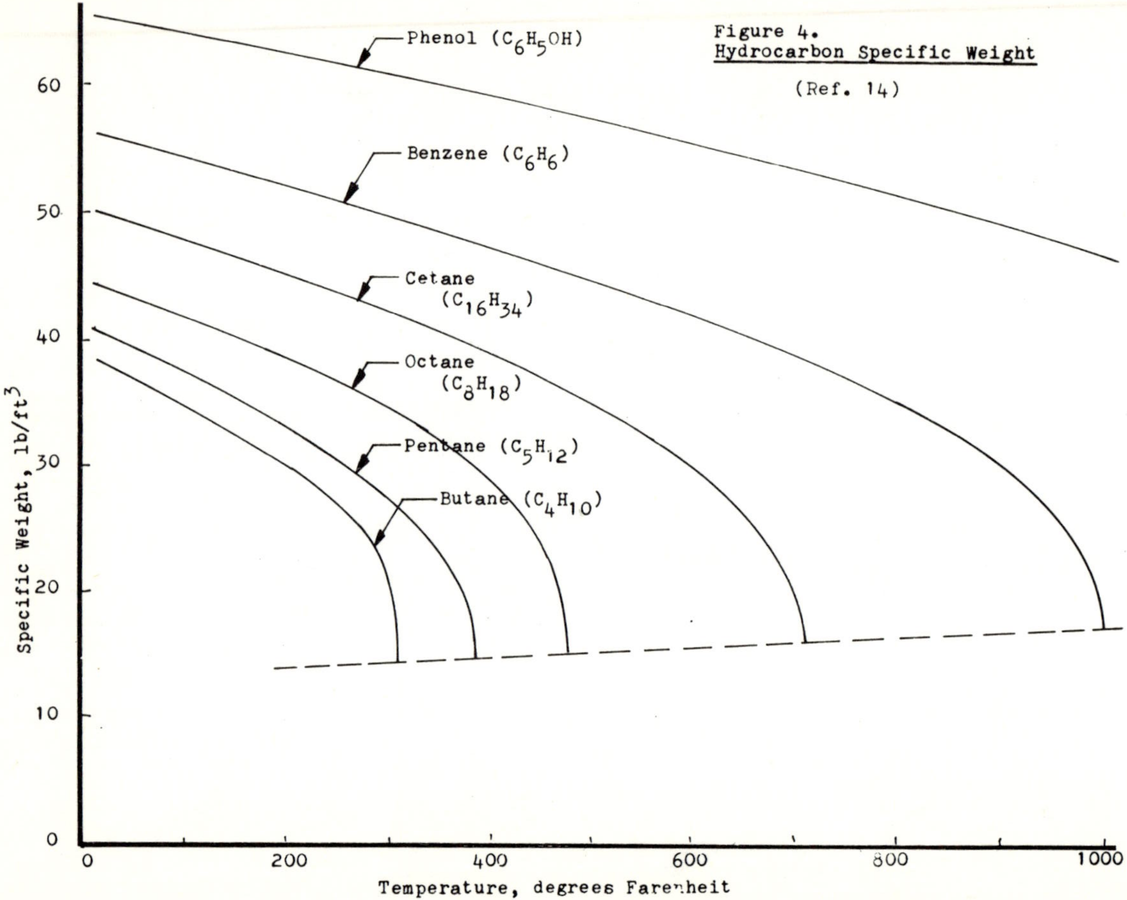

Figure 4.
<u>Hydrocarbon Specific Weight</u>

(Ref. 14)

Figure 5.
Water Solubility in Aviation Kerosene

(Ref. 16)

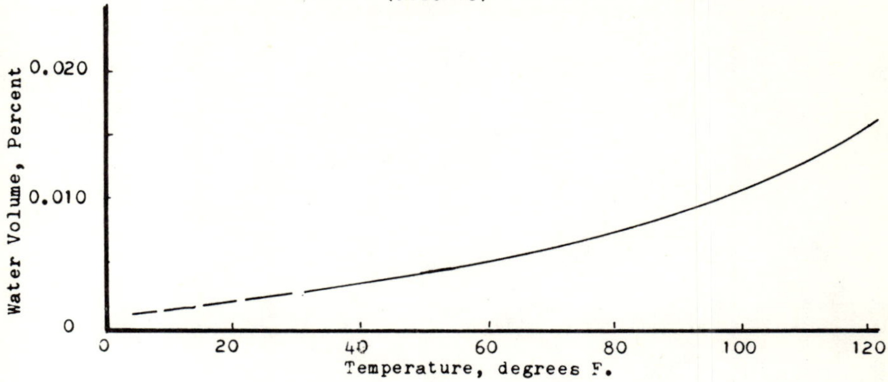

Figure 6.
Gas Solubility in Aviation Kerosene

(Ref. 17)

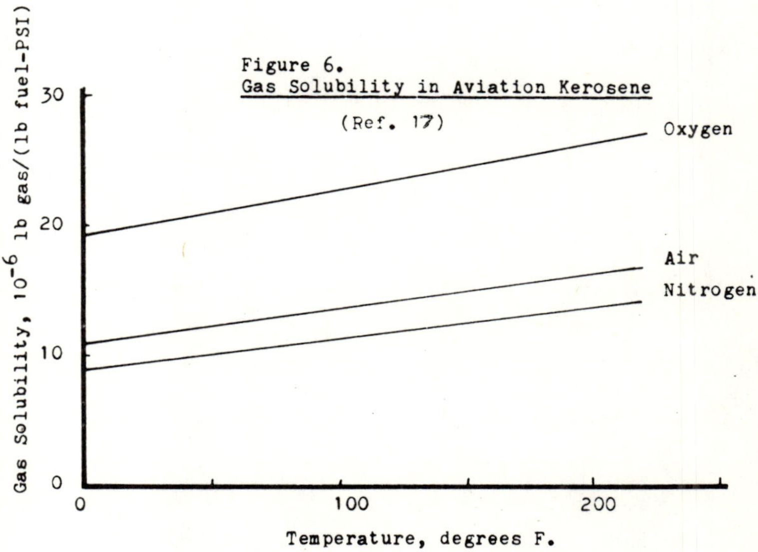

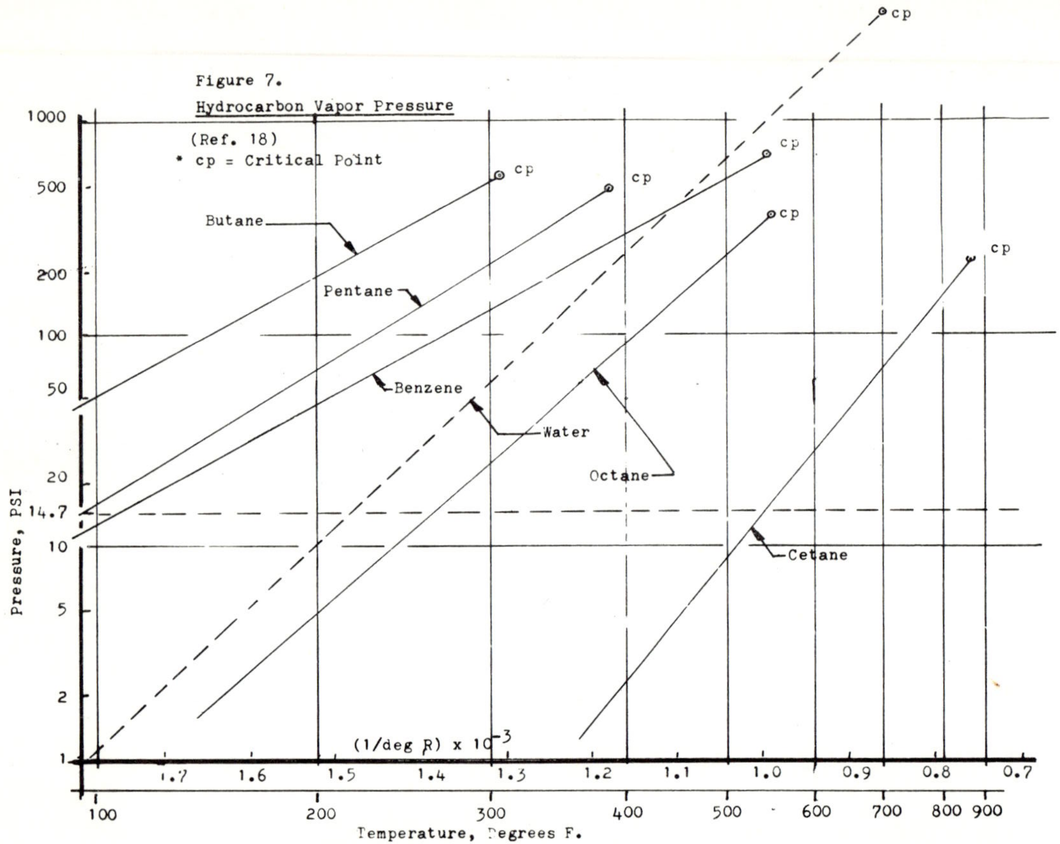

Figure 7.
Hydrocarbon Vapor Pressure

NOTES

NOTES

NOTES

NOTES